Home SOS

RGS-IBG Book Series

For further information about the series and a full list of published and forthcoming titles please visit www.rgsbookseries.com.

Published

Home SOS

Gender, Violence, and Survival in Crisis Ordinary Cambodia

Katherine Brickell

WILEY

Registered Offices
John Wiley & Sons, Inc., 111 River Street, Hoboken, NJ 07030, USA
John Wiley & Sons Ltd, The Atrium, Southern Gate, Chichester, West Sussex, PO19 8SQ, UK

Editorial Office
9600 Garsington Road, Oxford, OX4 2DQ, UK

For details of our global editorial offices, customer services, and more information about Wiley products visit us at www.wiley.com.

Wiley also publishes its books in a variety of electronic formats and by print-on-demand. Some content that appears in standard print versions of this book may not be available in other formats.

Library of Congress Cataloging-in-Publication data applied for

9781118898321 (hardback), 9781118898352 (paperback), 9781118898437 (ePDF)

Cover Design: Wiley
Cover Image: © Erika Piñeros

Set in 10/12pt Plantin by SPi Global, Pondicherry, India
Printed and bound in Singapore by Markono Print Media Pte Ltd

The information, practices and views in this book are those of the author and do not necessarily reflect the opinion of the Royal Geographical Society (with IBG).

10 9 8 7 6 5 4 3 2 1

For Christian and Stefan

Contents

List of Figures

List of Abbreviations

ADB	Asian Development Bank
ADHOC	The Cambodian Human Rights and Development Association
APRODEV	Association of World Council of Churches related Development Organisations in Europe
ASEAN	Association of Southeast Asian Nations
AusAID	Australian Aid
BRI	Belt and Road Initiative
BKL	Boeung Kak Lake
CAMBOW	The Cambodian Committee for Women
CCHR	The Cambodian Center for Human Rights
CDHS	Cambodia Demographic and Health Survey
CEDAW	The Committee on the Elimination of Discrimination Against Women
CNRP	Cambodian National Rescue Party
COHRE	Center on Housing Rights and Evictions
CPP	Cambodian Peoples' Party
CRP	Compliance Review Panel
DK	Democratic Kampuchea
DV	Domestic violence
ECCC	The Extraordinary Chambers in the Courts of Cambodia
ELC	Economic Land Concession
FDI	Foreign Direct Investment
FUNCINPEC	National United Front for an Independent, Neutral, Peaceful and Cooperative Cambodia
GADC	Gender and Development for Cambodia
GADNET	The Gender and Development Network
GIZ	Die Deutsche Gesellschaft für Internationale Zusammenarbeit
GMS	Greater Mekong Subregion

GTZ	Die Deutsche Gesellschaft für Technische Zusammenarbeit
ICNL	The International Center for Not-for-Profit Law
IDI	Inclusive Development International
ILO	International Labour Organization
LANGO	The Law on Associations and Non-Governmental Organizations
LICADHO	Cambodia League for the Promotion and Deference of Human Rights
LMAP	Land Management and Administrative Project
MOWA	Ministry of Women's Affairs
NAPVAW	The National Action Plan to Prevent Violence Against Women
NCSWF	The National Committee for Upholding Cambodian Social Morality, Women's and Khmer Family Values
NGO	Non-governmental organisation
NGO-CEDAW	Cambodian NGO Committee on CEDAW
NIS	National Institute of Statistics
OHCHR	Office of the United Nations High Commissioner for Human Rights
P4P	Partners for Prevention
PRK	People's Republic of Kampuchea
RGC	Royal Government of Cambodia
SNC	Supreme National Council
UN	United Nations
UNDP	United Nations Development Programme
UNGA	United Nations General Assembly
UNHCR	The Office of the United Nations High Commissioner for Refugees
UN-HABITAT	United Nations Human Settlements Programme
UNODC	United Nations Office on Drugs and Crime
UNTAC	United Nations Transitional Authority in Cambodia
UNWOMEN	The United Nations Entity for Gender Equality and the Empowerment of Women
US	United States
US$	United States Dollar
VOA	The Voice of America
WHO	World Health Organization
WMC	Women's Media Centre

About the Author

Katherine Brickell is Professor of Human Geography at Royal Holloway, University of London (RHUL), UK. Her research cross-cuts social, political, legal, and development geography, with a longstanding focus on the domestic sphere as a precarious and gendered space of contemporary everyday life. She has over 15 years of research experience in Cambodia and since 2017 has begun to undertake new collaborative work in the UK and Ireland. *Home SOS* is Katherine's first monograph and follows the publication of co-edited collections including *Translocal Geographies* (2011 with Ayona Datta), *Geographies of Forced Eviction* (2017 with Melissa Fernández Arrigoitia and Alex Vasudevan), *The Handbook of Contemporary Cambodia* (2017 with Simon Springer), and *The Handbook of Displacement* (2020 with her RHUL colleagues). In recognition of research excellence, she was conferred the Gill Memorial Award by the Royal Geographical Society (RGS-IBG) in 2014 and the Philip Leverhulme Prize in 2016. She is editor of the journal *Gender, Place and Culture* and is former Chair of the RGS-IBG Gender and Feminist Geographies Research Group.

Series Editor's Preface

The RGS-IBG Book Series only publishes work of the highest international standing. Its emphasis is on distinctive new developments in human and physical geography, although it is also open to contributions from cognate disciplines whose interests overlap with those of geographers. The Series places strong emphasis on theoretically informed and empirically strong texts. Reflecting the vibrant and diverse theoretical and empirical agendas that characterize the contemporary discipline, contributions are expected to inform, challenge and stimulate the reader. Overall, the RGS-IBG Book Series seeks to promote scholarly publications that leave an intellectual mark and change the way readers think about particular issues, methods or theories.

For details on how to submit a proposal please visit:
www.rgsbookseries.com.

David Featherstone
University of Glasgow, UK
RGS-IBG Book Series Editor

Acknowledgements

It is difficult to know where to start writing these acknowledgements. I first submitted the proposal for *Home SOS* in 2012, and eight years on, its journey into print has finally come to an end. This end point has only been reached through sustained, and much appreciated, professional and personal support.

The book would not exist without the time, generosity and emotional energy of participants in sharing their stories of domestic life in Cambodia. It has been an honour and a privilege to listen to and write about their intimate experiences in *Home SOS*. The four studies the book is based on have been made possible by the interpreters and research assistants I have worked with – young and inspiring Cambodians who I am incredibly grateful to for their dedication and kindness. I feel saddened and torn that I cannot name them here given the political sensitivities of the book, which have only intensified over the course of writing it. I am also grateful to the many photographers who have allowed me to use their images in the book to provide the reader with a visual sense of home precarities unfolding in Cambodia. The joint reporting of Cambodian and international journalists on forced eviction in national newspapers, now shut down or under new management, has been particularly helpful to understanding the frequency and impact of women's activism in relation to Boeung Kak Lake.

Thank you to the RGS-IBG Book Series for your expertise and understanding in bringing the book to fruition over such a long period of time. Thank you to Neil Coe and Dave Featherstone for providing constructive feedback at each stage, and to Jacqueline Scott for liaising with me for so many years.

The Leverhulme Trust has been instrumental in enabling the time to write the monograph. I would also like to thank the funders of the research, the Economic and Social Research Council (ESRC) and the Royal Geographical Society. The PhD research that forms the initial basis for *Home SOS* was supervised by Sylvia Chant at the London School of Economics. Thank you Sylvia for giving me the best start in my academic journey, for believing in me and for showing me what passion and drive can achieve. Since 2008 I have been a researcher at Royal

Holloway, University of London and having been home to my scholarly endeavours for over a decade, it is only fitting that I thank my colleagues, both academic
and administrative, for the support offered.

In the academic community, I am also honoured to have had ongoing support
from geographers who have read and commented on the many iterations of the
book. Ruth Craggs is of especial note for having read drafts of each and every
chapter, on multiple occasions. Since meeting for the first time at the Las Vegas
AAG in 2009, I have rarely felt lonely in academia because of our friendship and
our writing side-by-side across cafés in London. I am also grateful for the many
writing retreats we have been on, memorably battling through snow to get there,
and taking trips to garden centres as shared spaces of happiness in which to clear
our heads. Writing retreats have been a key way I have managed to push the book
substantively forward. Thank you Harriet Hawkins for our cherished writing
retreats together, and for being such a positive and reassuring figure in the journey
of this book. I have also benefitted from the insightful feedback given to me
by James Tyner, Nithya Natarajan, and Laurie Parsons, which extended the
book's ambitions in the final year of its writing. The opportunity for honed
thinking has also been facilitated by the feedback shared with me through
departmental seminars at the University of Leicester, King's College London
and Durham University.

The long journey of the book's coming to pass has arisen through personal
circumstances that I could never have predicted when I began writing. Soon after
returning from maternity leave in 2015 I was diagnosed with a rare cancer,
Pseudomyxoma peritonei (PMP), and took medical leave to undertake major
surgical and chemotherapy treatment. As the disease is so rare, it is important
that I use this opportunity to raise awareness of it (see Macmillan and Cancer
Research UK web pages). Throughout my treatment, and in the years since,
I have received practical and emotional advice from the PMP community of fellow survivors and its organisation run by carers and patients (https://www.
pseudomyxomasurvivor.org). My being here is testament to the NHS and the
dedicated surgeons and nurses at the Peritoneal Malignancy Institute
at Basingstoke, who I want to sincerely thank. I would like to note Mr Sanjay
Dayal, my lead consultant surgeon, and specialist nurse Vicki Pleavin-Evans for
being there, still, at the end of the phone with your wise words. Given the significance of the treatment, I would also like to thank Gary Walker and Crystal Sutar
at Grafton Tennis Club for working with me slowly, but surely, every week to
build my confidence and trust in my body again.

Home SOS has been a monograph that has been with me on this unexpected
journey, offering a sense of continuity and reflection in difficult times. Thank you
Ruth Jacob, Ali Moss and Jana Ulph for providing me with a safe space to offload
and to laugh; to Ellen Wiles for inspiring me and offering solidarity; and to
Christine Widerøe Frenvik for our enduring friendship, which began upon a
chance meeting on the streets of Siem Reap so many moons ago. Finally, I would

like to warmly thank my family, without whom none of this would have been possible. I owe a huge debt of gratitude to my parents and sister for caring for me through years of fieldwork, and offering practical, childcare and emotional support when it mattered most. This book is dedicated to my husband Christian and son Stefan, the loves of my life, from whom I have gained daily encouragement and joy. You have steadfastly held my hand, through my concurrent health challenges and the writing of *Home SOS*. I simply cannot thank you enough.

Katherine Brickell
Department of Geography, Royal Holloway, University of London,
Egham

Chapter One
Fire in the House

Introduction

Home SOS casts a vital spotlight on the domestic sphere as a critical, yet over-looked, vantage point for understanding the trajectory of Cambodia, a Southeast Asian nation known for its encounters with genocide (Hinton 2005; Kiernan 2002), reconciliation and peacebuilding (Ciorciari and Heindel 2014; Gidley 2019; Hughes and Elander 2017; Öjendal and Ou 2013, 2015; Peou 2007, 2018; Richmond and Franks 2007), post-conflict transition (Öjendal and Lilja 2009; Peou 2000), and economic transformation (Hughes and Un 2011; Hughes 2003; Springer 2010, 2015). During Pol Pot's genocidal reign (1975–1979), the home was rendered 'ground zero' in efforts to break apart families, intimacies and other relations of the former society. As a lifeworld 'constituted by relatively stable associations, relatively known and shared histories' (Bhabha 1994, p. 42), it had to be destroyed. Cambodia needed to be 'killed', to cease to exist both literally and symbolically, for the radical communist revolution to be built (Tyner 2008, p. 119).

After decades of upheaval and displacement, Cambodia moved through a so-called 'triple transition' (Peou 2001, p. xx): from armed conflict to peace, from political authoritarianism to liberal democracy, and from a socialist economic system to a market-driven capitalist one. The accelerated neoliberal path that the country has taken, and its unravelling experimentation with democracy, has had a profound influence on domestic life. In this political

Home SOS: Gender, Violence, and Survival in Crisis Ordinary Cambodia, First Edition. Katherine Brickell.
© 2020 Royal Geographical Society (with the Institute of British Geographers). Published 2020 by John Wiley & Sons Ltd.

economy context, at a time of formal peace, *Home SOS* contends that home is, once again, a spatial epicentre of intimate violences, thought to have been consigned to its genocidal past. It is not only that wounds remain in dysfunctional marriages and traumas carried through time, but rather, with its formal end, Cambodia opened up to the world, and new wounds have reconfigured and compounded the old.

The dual focus of *Home SOS* on domestic violence and forced eviction shows that the home is not a 'pre-political' or 'unexceptional' space (Enloe 2011, p. 447) separate from these political changes. Its internal intimacies and external-facing dynamics have the capacity to temper, retell, and rework the grand narratives of change that have so dominated writing on Cambodia. As a pivotal space of bionecropolitical world (un)making, the home demands greater scrutiny. It is where the production and destruction of life is perhaps most regularly and intensely expressed, yet it is systematically overlooked in theoretical writing in geography and related disciplines.

The starting point of the book then is the 'extra-domestic' home, in and through which, multiple forms of violence flow and coalesce. This 'extra-domestic' reading derives 'precisely from the fact that it [the home] had always in one way or another been open; constructed out of movement, communication, social relations which always stretched beyond it' (Massey 1994, p. 171). The home is host to violences that are played out through the microdramas of daily life, but also through public political worlds that influence, and are influenced by, the domestic (Blunt and Dowling 2006; Brickell 2012a, 2012b; Nowicki 2018). Tracing these violences through the realm of the extra-domestic problematises the narrowness of 'crisis-affected' and 'crisis-prone' descriptors limited to countries and regions experiencing war, conflict, and natural disasters. The intimate wars of domestic violence and forced eviction render the home a crisis-affected and crisis-prone space, both inside and outside, of these formal hostilities and calamities. *Home SOS* thus works to reaffirm and reprioritise the home as a political entity that is foundational to the concerns of human geography.

More specifically, the book contributes to geographical scholarship on intimate geographies of violence and crisis that have not commanded as much attention as high-profile public ones. The concern of *Home SOS* is not simply the enactment of domestic violence and forced eviction but how they are experienced and contested. While neither domestic violence nor forced eviction are formal emergencies, the title of the book deploys the SOS distress signal. This internationally known code has been used by forced eviction women activists in Cambodia as a political statement of emergency to oppose their dispossession (Figure 1.1). Their public SOS calls reassert the importance of imbuing everyday structural violences with a sense of scholarly and political urgency that is so often lost, given the seeming intransigence of routine violences (Philo 2017; Scheper-Hughes 1996).

Figure 1.1   SOS sign in Boeung Kak Lake, Phnom Penh, 2011. Source: © Ben Woods. Reproduced with permission.

Charting the journey of Cambodia's first-ever domestic violence law (DV law), alongside women's housing activism against forced eviction, I examine how each usher women's bodily presences from the home into the wider world to make life more liveable. They are, on the face of it at least, refusals for the injustices they embody to be known only in the private realm. As I go on to demonstrate, however, the political economy of Cambodia has conspired to stymy, and in some cases quash, the transformative potential that each might hold.

Based on over 300 interviews, conducted over the past 16 years, the research presented in *Home SOS* pivots around experiences of intimate violence and the work of survival as told through the gendered stories of 'ever-married' Cambodian women.[1] It incorporates continuums and rearticulations of violence from the Khmer Rouge period and includes discussion of episodic shocks and endemic precarities played out in the lifeworlds of participants in the four decades since. In what follows, I explore the ways in which home and intimate life are sustained, contested, and disrupted through marital relationships 'in crisis' and which are lived out, and shaped through, this history and newer manifestations of crisis taking place.[2] The book therefore aims to document, make visible, and interpret women's experiences of violence and survival in, and through, the home as an irreplaceable centre of significance in still challenging times. Looking to the home as the spatial epicentre of married life, the book contributes to the now growing body of research in geography on family and intimate relations (see Tarrant and Hall 2019 for a recent review). While noting that home and family are not necessarily coterminous, and intimate relations are not only marital but take a variety of forms, I take the marital home as my central referent point. A little over a decade ago, the family was considered an absent presence in the discipline (Valentine 2008), and marriage as the 'commonest form of social cooperation' had been neglected in the social sciences on the grounds of its ubiquity, taken for granted, and more-than-individual character (Harker 2012; Jackson 2012). By showing how marriage is a foremost social relationship through which domestic violence and forced eviction are articulated, *Home SOS* further questions the validity of these grounds for dismissal.

In Cambodia 'family is at the core of society' and is traditionally headed by a man who is invested with meeting its economic and social needs, and women its housekeeping and child rearing responsibilities (Ministry of Womens' Affairs (MOWA) 2015, p. 23). It is normatively accepted that Cambodian wives must provide 'shade' including 'shelter, safety and prosperity' for their children (Kent 2011a, p. 197–198) and are responsible for the management of conflict in family life. Women in Cambodia have primary duty for the social reproduction of their households and, in turn, they gain important public value through domesticity. Nineteenth century normative Cambodian poems such as the *Chbab Srei* (Rules for Women) refer derogatorily to women who walk too loudly in the house as to make it tremble, and should a woman act too forcefully, neglect her household responsibilities, or not fulfil the entirety of her husband's demands, then blame can be assigned to her for the breakdown of their marriage. Divorced women are typically regarded as socially incomplete in Cambodian society (Ovesen, Trankell, and Öjendal 1996) and a Khmer proverb reminds daughters that, 'you should be married before you are called an old maid' (cited in Heuveline and Poch 2006, p. 102). Women's homes and bodies therefore risk being considered as lacking when marriages fail. To avoid this, women in Cambodia are invested with the demands of ensuring the harmony, intimacy, and warmth of their marriages and wider family in the home. Domestic violence and forced eviction not only challenge women's capacity to fulfil these expectations, but they also threaten the material and symbolic foundations of everyday life from which life is grown and sustained. Both domestic violence and forced eviction have, therefore, a disproportionate impact upon women given the intensities of homemaking in comparison to men.

The twin focus of *Home SOS* on domestic violence and forced eviction arises, in part too, from their status as human rights abuses, which witness the 'mutual absorption of violence and the ordinary' (Das 2007, p. 7). The home is, perhaps, the most ordinary of spaces on which geographers train their research and analysis. An intimacy sustaining vision of the 'ordinary' home, however, is complicated by domestic violence and forced eviction. which both disrupt the safety, stability, and comfort that the space should normatively afford. As Lauren Berlant (1998, p. 281) argues, intimacy … involves an aspiration for a narrative about something shared, a story about both oneself and others that will turn out in a particular way. Usually, this story is set within zones of familiarity and comfort: friendship, the couple, and the family form, animated by expressive and emancipating kinds of love … This view of 'a life' that unfolds intact within the intimate sphere represses, of course, another fact about it: the unavoidable troubles, the distractions and disruptions that make things turn out in unpredicted scenarios.

Berlant's conception of intimacy brings to the fore the 'unpredicted scenarios', which are perceptible and woven into the ordinary. Under this guise, the 'ordinary' cannot be equated with the benign and harmless (Mayblin, Wake, and Kazemi 2019); its study requires that scholars go beyond and interrogate the taken-for-granted. The potential for such discordance, ambiguity, and negativity in experiences of ordinary life has been a motivating focus in the critical geographies of the home sub-field, which emerged in the mid-2000s (Blunt and Varley 2004; Blunt and Dowling 2006). As Jeanne Moore (2000, p. 213) argued, there was a 'need to focus on the ways in which home disappoints, aggravates, neglects, confines and contradicts as much as it inspires and comforts'. Decades on, there is room to think more about how to theorise the home as a space that absorbs as well as repels 'trouble'. The fact that in the family geographies sub-field 'everyday processes of family conflict, trouble and disruption' currently only 'occupy a marginal space in the geographic literature' (Tarrant and Hall 2019, p. 4) lends further weight to this task.

In *Home SOS*, I take the 'crisis ordinary' (Berlant 2011) as the lead conceptual frame through which I deepen theoretical engagements on intimate geographies of violence in these sub-fields, and geography more broadly. Domestic violence and forced eviction, I contend, are both instantiations of the crisis ordinary, 'when the ordinary becomes a landfill for overwhelming and impending crises of life-building and expectation whose sheer volume so threatens what it has meant to 'have a life' that adjustment seems like an accomplishment' (Berlant 2011, p. 3). *Home SOS* explores this social theory through a grounded, embodied, and long-running account of crisis ordinaries that are unfolding in Cambodia and that allow elite men, in particular, to maintain the balance of power at the expense of women and their homes.

The book that follows is an ambitious, expressly experimental one, which rather being a single-issue monograph is dedicated to the integrated study of domestic violence and forced eviction. Domestic violence and forced eviction are typically understood as discrete and unconnected violences, the former usually discussed as a type of interpersonal violence and the latter as a 'violence of property' (Springer 2013a, p. 608). However, a core argument made in the book is that the intimate injustices and precarious journeys that domestic violence and forced eviction embody are also intimately linked and, in some instances, simultaneously faced in women's lives. I read and weave these interconnections through hybrid yet complementary theoretical engagements with the crisis ordinary and survival-work, bio-necropolitics and precarity, intimate war and slow violence, law and lawfare, and rights to dwell (see Chapter 2). Drawing on literatures from development studies, women's studies, anthropology, politics, and international relations (IR), and spanning development, legal, political, and social geography, I provide, ultimately, an original book-length study that combines domestic violence and forced eviction to offer relational insights between these violences.

In this introductory chapter I set up some of the major and interconnected arguments that the book takes forward: first, that domestic life is lived and manifest in and through crisis ordinaries; second, that domestic violence and forced eviction necessitate what I call 'survival-work'; and, third, that they are both forms of gender-based violence. I then turn to the research trajectories that led to the convergence of the domestic violence and forced eviction research and material. I also include reflection on key overarching methodological approaches taken and how these are reflected in the content and ethos of *Home SOS*. Finally, I move on to the structure of the book and summarise its findings.

Crisis Ordinaries of Domestic Violence and Forced Eviction

If Berlant's work on the crisis ordinary reminds scholars 'to make sense of the ways in which subjects find themselves habituating, situating, desiring, or feeling in the world, day to day, often amid conditions of cruelty' (Cram 2014, p. 374), then *Home SOS* shows how systemic crises are entrenched in the home and speak to broader patterns of violences as part and parcel of the vernacular landscape of Cambodia. Both domestic violence and forced eviction can be considered as spatial exclusions or 'expulsions' (Sassen 2014) from home and living space as part of a wider diagnosis of unstable and disconcerting times. Yet they are forms of ongoing loss that a language of expulsion risks eliding. Taken together, domestic violence and forced eviction work to underscore the significance of 'less sensational, yet nevertheless devastating' dislocations from home (Vaz-Jones 2018, p. 711). They are singular yet inter-linked forms of crisis ordinariness rooted in metaphorical and/or physical displacement. Above all, however, they are lived experiences, which the book prioritises. Just as domestic violence can lead to forced eviction from home, empirical data in the book attests to the emergent links that can be drawn between forced eviction, domestic violence, and marital breakdown. In order to start building a holistic understanding of intimate violences encountered by women in contemporary Cambodian society, the equivalences and morphisms that exist between domestic violence and forced eviction are explored in the pages that follow.

Doing so aims to counteract how 'there are many forms of precarity and disorder not captured under the designation of "crisis" that do not command the same critical attention' as high profile disasters (Ahmann 2018, p. 147). In this regard, the book thus builds from recently published work in two areas. The first relates to crisis geographies that are multi-scalar and embedded in political economy dynamics (such as austerity), but are experienced intensely at the personal scale and articulated through everyday fragilities of family life (Hall 2019). In *Home SOS*, for example, I show how marriage is an intimate yet precarious union that is influenced by wider political structures and processes, including Cambodia's recent history of humanitarian crisis at the hands of the Khmer Rouge. A second area that *Home SOS* builds from is the vital importance of 'non-eventful geographies' of violence and everyday life – violences that are too often left out of social science work on suffering and dying (Wilkinson and Ortega-Alcázar 2019). Such 'quasi-events', Elizabeth Povinelli (2011, p. 13) argues, 'never quite achieve the status of having occurred or taken place'. My ambition, then, is to reposition domestic violence and forced eviction as events, rather than quasi-events, that happen, and are happening, in the extra-domestic realm.

Domestic life for many women in Cambodia, I show, is saturated with crises, which are viewed in ordinary and anticipatory terms. The Cambodian Buddhist expression 'fire in the house' embodies the idea, for example, that in order to maintain a harmonious household, women are responsible for suppressing three fires of potential conflict within the home – parents, husbands, and 'others' (Derks 2008). Often uttered in relation to domestic violence, the Khmer proverb 'Plates in a Basket will Rattle' speaks to the saturation and management of conflict (fire) in everyday life that is normal rather than diversionary. Crisis in this guise can be approached as a 'prosaic', 'the routinization of a register of improvisations lived as such by people and, in this sense, belonging at most to the domain of the obvious or self-evident, and at least to the banal or that which no longer evokes surprise' (Mbembé and Roitman 1995, p. 326). The first-ever survey of domestic violence in Cambodia opens with a vignette that evokes this further: 'if people live in the same house there will inevitably be some collisions. It's normal … it can't be helped. But, from time to time, plates break. So do women' (Zimmerman 1995, np). Through the life stories of female respondents, *Home SOS* focuses on these collisions and, in some tragic cases, their fatalities; my argument is that women's homes and bodies are being broken, time and again, in the making of Cambodia today.

The joint study of domestic violence and forced eviction reveals how this pernicious breaking exceeds baseline expectations of domestic conflict. Not only this, but the home is a dynamic space of potentiality (Povinelli 2011) in which women are innovating under these debilitating circumstances in different ways. Some women I spent time with continued to manage the flames, others felt compelled to ignite them in the public sphere to try and extinguish their threat, whilst others exited marriage to escape the heat altogether. In the ordinary, then, it is possible to read the eventful, the memorable, and also the episodic, 'occasions that make experiences while not changing much of anything' (Berlant 2007, p. 760). As Susan Fraiman (2017, p. 123) writes in relation to the reproduction of the ordinary, it is women who 'generally get the brunt' of this work. Women's firefighting takes many forms in contemporary Cambodia and the book ignites discussion of domestic violence and forced eviction as fires with multiple and politically imbued sources, responses, and outcomes.

Home SOS goes on to show, for instance, how the mundanity of the crisis ordinary continues as marginalisation and containment of these supposed 'non-events' to the home by a government unwilling to tolerate its spilling out into public and the concomitant political questions and challenges to its power that this may bring. Government attempts to keep disruption to the established order at a minimum enables the continued production of death, social or actual, through de facto gerrymandering whereby political advantage is achieved by manipulating and regulating the boundaries of home and the 'fire' within. Women have the right to dwell free from violence, but in their everyday lives, and in their pursuit of this goal, are subject to a bio-necropolitical brutality that the book brings into view through its joint focus on domestic violence and forced eviction.

The Survival-Work of Domestic Violence and Forced Eviction

To examine women's injuries, but also the survival practices, that are performed in the domestic domain, I synergistically place domestic violence and forced eviction within an expansive conceptualisation of work that exposes capitalist patriarchy as the requirement that women perform survival-work across private and public realms (Dalla Costa 1972; Mies 1982). The centrality of social reproduction to accounts of violence and dispossession thus take on critical importance (Fernandez 2018). As Maria Mies (2014, p. 2) elaborates in relation to the accumulation of wealth, productive capital, and control by men:

> Today, it is more than evident that the accumulation process itself destroys the core of the human essence everywhere, because it is based on the destruction of women's autarky over their lives and bodies.

As feminist writing has long set out, women's labour in the home is not a separate social sphere located outside of economic relations but is integral to it. Despite this, 'housework is not counted as work, and is still not considered by many as "real work"' (Federici 2012, p. 127). My reference to survival *work* therefore tries not to lose sight of the range of labour that women undertake in the extra-domestic home, but which remains largely invisible in writings on political and economic change in Cambodia. The home is a discursive construct and product of ongoing and demanding labour that is intimately connected, rather than sealed off, from political meaning and impact. Just as Sylvia Chant (2007) identifies a 'feminisation of responsibility and obligation' taking hold in the Global South as rising numbers of poor women of all ages are working outside the home and are continuing to perform the bulk of unpaid reproductive tasks, *Home SOS* adds the precarities of domestic violence and forced eviction to this growing suite of duties. As such, it directly responds to Berlant's (2007, p. 757) incitement that 'we need better ways to talk about activity oriented toward the reproduction of ordinary life: the burdens of compelled will that exhaust people taken up by managing contemporary labor and household pressures'.

Evidence from around the world demonstrates that it is women who, with deepening inequalities, are shouldering the burden of adjustment as 'shock absorbers' and carers for households on the edge of survival (Brickell and Chant 2010; Cappellini, Marilli, and Parsons 2014; Elson 2002; Gill and Orgad 2018; McDowell 2016; Sou and Webber 2019). Given the corporeal and material hardships of domestic violence and forced eviction, as well as the dramatically limited and high-stake choices that they both entail, the significance of emotional (Hochschild 1983) as well as physical labour cannot be discounted. Focusing on daily spaces and routine situations reveals how 'precarity is embedded in the mundane tasks of the domestic, and as a result, unevenly impacts women whose traditional roles as mothers and caretakers mean that they are often at the fore of place-making practices and responsibilities' (Muñoz 2018, p. 411). Both domestic violence and forced eviction are traumatic ruptures in time and space of domestic and social reproduction in all their dimensions. This includes the symbolic dimensions of identity and representation (Meertens and Segura-Escobar 1996) that Cambodian women are typically responsible for in their familial lives.

If patriarchy is to be understood as men's violent appropriation of women's labour as a dominant force of production (Federici 2014, p. x), then the survival-work that women perform is being co-opted as a means to uphold the viability of the Cambodian state. This is because 'the reproduction of human beings is the foundation of every economic and political system' and 'the immense amount of paid and unpaid domestic work done by women in the home is what keeps the world moving' (Federici 2012, p. 2). In other words, the Cambodian government's accumulation model is being strengthened not only through the waged labour of women in the garment factories sustaining its economic growth but also the unwaged labour that women are undertaking in the home. Both are 'productive' for the reproduction of 'big men' and the modern Cambodian state; predicated on the control of women's homes and bodies that they desire to have firm and lasting dominance over. The home and women's labour in it, are cornerstones upon which capital accumulation is forged. Cindi Katz (2001, p. 709) writes that, as a result, it is important to critically study 'the material social practices through which people reproduce themselves on a daily and generational basis and through which the social relations and material bases of capitalism are renewed' (see also Federici 2018a). The survival-work entailed in living with, or on, from domestic violence and forced eviction pushes the importance, therefore, of 'broadening the concept of labor to more fully articulate the dialectics of production and reproduction' (Tyner 2019, p. 1307; see also Rioux 2015).

With these points in mind, a mainstay argument pursued in *Home SOS* is that women are not only disproportionally impacted upon by domestic violence and forced eviction as events and processes in themselves but that they are also tasked with the adjustive work of homemaking and the management of everyday precarity stemming from this. The expectation of, and necessity for, women's survival-work in the crisis ordinary is highly problematic and the home is a key setting where this can be identified and studied. Homemaking is a cultural process and since a 'culture only exists as a sum total of its iterations' (Macgregor Wise 2000, p. 310), this 'pragmatic (life-making)' and 'accretive (life-building)' work becomes endangered and more arduous when the conditions for predictability are undermined (Berlant 2007, p. 757). My reference to survival is therefore a 'back to basics' move (Heynen 2006, p. 920), which recognises the centrality of home to survival and the labours that go into the meeting of basic material need, familiarity, and comfort. 'Practices of survivability', Loretta Lees, Annunziata, and Rivas-Alonso (2018, p. 349) write, include 'all of the different practices people employ to stay put' and that counteract 'blanket statements of neoliberal hegemony'.

The survival-work that women do to 'stay put' is underpinned by the home being both 'a nodal point of concrete social relations' and 'a conceptual or discursive space of identification' (Rapport and Dawson 1998, p. 17). The home as a physical and ideological entity reflects the significance of place in 'dis*place*-ment' (Davidson 2009, p. 226, emphasis in original). As I conjectured earlier, domestic violence and forced eviction are attempted deprivations of home, its material infrastructure, and/or sense of belonging. Domestic violence is one of the most overlooked forms of displacement as women are often forced to leave their homes suddenly, without their possessions, to an unknown and unfamiliar place (Warrington 2001; Graham and Brickell 2019). In the United Kingdom, for example, tens of thousands of forced migrations occur each year as a result of domestic violence yet it is a country designated as having no internally displaced persons (Bowstead 2015, 2017). Such mobilities 'could certainly be understood *as* emergency' given that they 'demand highly intensive forms of movement that radically transform one's life chances and quality of life' (Adey 2016, p. 32). Women experience 'a process of spatial churn' as they undertake individual, isolated journeys, and move multiple times before they are able to find a settled home (Bowstead 2015, p. 317). For women who cannot leave meanwhile, they may feel 'homeless at home' (Wardhaugh 1999), living in, and managing, a violent environment through daily and multiple forms of survival-work. The ideological scripting of home as intimate and safe can also lead to women tolerating violence so as not to signal a deep failure or collapse of home (Price 2002). The book shows that in Cambodia, this ideological scripting is strongly focused on women's familial responsibilities and has a similar influence. Part of what makes the violence so untenable and cruel is the survival-work required to keep up this pretence.

The threat of forced eviction can also lead to the chipping away of home, materially and metaphorically, and the eventual displacement and dislocation of families ejected from their homes leads to their 're-settlement'/homelessness. Much like the 'spatial churn' that domestic violence survivors can face, forced eviction also produces a harmful turbulence that women must typically counteract through their emotional and physical labours. Survival-work inside, and outside of, the home is therefore an important, yet still neglected, part of labour and economic geography that warrants scholarly development (Strauss 2018).

The Gender-Based Violences of Domestic Violence and Forced Eviction

While domestic violence and forced eviction are significant issues for women in their own right, *Home SOS* contends that, taken together, they most clearly and definitively demonstrate the gendered contingency of existence in contemporary Cambodia. Situating the book as a feminist geography intervention with broader (inter)disciplinary significance, I venture that both are acts of gender-based violence. Gender-based violence is defined as 'violence which is directed against a woman because she is a woman or that affects women disproportionately' (Committee on the Elimination of Discrimination Against Women (CEDAW) 1992, np). This includes acts that inflict physical, mental, or sexual harm or suffering, the threat of such acts, coercion, and other deprivations of liberty (CEDAW 1992).

Domestic violence is a well-known form of gender-based violence that encompasses violence against women by an intimate partner and/or other family members, wherever this takes place, and in whatever form, be this physical, sexual, psychological, or economic. Adult women account for the vast majority of domestic violence victims globally. The insidious nature of domestic violence in Cambodia was first highlighted by inaugural research conducted in the mid-1990s by the Project Against Domestic Violence with 1374 women (Nelson and Zimmerman 1996). The study found that 16% of women surveyed reported physical abuse by their spouses and 8% acquired injuries mostly to the head. More recently, a nationally representative Partners for Prevention (P4P) (2013) study showed that one in four women (25.3%) in Cambodia had experienced in their lifetime at least one act of physical or sexual violence, or both, perpetrated by an intimate partner. Figures published in the country's *National Survey on Women's Health and Life Experiences* (MOWA 2015) report too that approximately one in five women (between 15 and 64 years old) who had ever been in a relationship had experienced physical and/or sexual violence by an intimate partner at least once in their lifetime. Further to this, three out of four women who had experienced physical and/or sexual violence had encountered severe acts of violence that were likely to cause injury. These include being hit with a fist or something that could hurt; dragged, kicked, or beaten up; threatened or hurt by a gun, knife, or other weapon. The survey also found that women are much more likely to experience frequent acts of violence rather than one-off incidents. Domestic violence occurrence is not an isolated event, but is a pattern of violative behaviour that women encounter. This includes emotional abuse, with almost one in three (32%) of 'ever-partnered' women aged 15–64 in Cambodia reporting controlling behaviour and/or the threat of harm (MOWA 2015).

In contrast to domestic violence, it is unusual that forced eviction is explicitly framed as gender-based violence. Forced eviction is viewed as a widespread and systematic human rights violation in Cambodia but is rarely discussed as gender-based violence. Cambodia is infamous for the scale and brutality of forced evictions occurring in, and beyond, Phnom Penh under the auspices of development. According to WITNESS (2017), at least 30 000 residents of the capital city, Phnom Penh, have been forcibly evicted, and approximately 150 000 Cambodians throughout the country are at risk of forced eviction. Underscoring the magnitude of the issue, between January 2000 and March 2014, the human rights LICADHO (2014a) documented more than 500 000 Cambodians affected by state-involved land conflicts in investigations covering roughly half the country.

Kaori Izumi's (2007, p. 12) writing on Southern and East Africa was, until recently, rare in its direct contention that forced eviction 'represents a form of gender-based violence in itself, as well as often being accompanied by other acts of extreme violence against women, including physical abuse, harassment, and intimidation, in violation of women's human rights'. A United Nations (2014a, p. 16) fact sheet on forced eviction now acknowledges that, 'women often tend to be disproportionately affected and bear the brunt of abuse during forced evictions'. Furthermore, it affirms that forced eviction commonly entails 'direct and indirect violence against women before, during and after the event'.[3] Forced eviction hurts in multiple senses, but is particularly traumatic for women given its targeting of the domestic sphere. As Yorm Bopha, an interviewee activist evocatively told me,

> My message is that home is life for women. Even the bird needs a nest. Even corpses need a cemetery. The most important place is home. If we do not have a home, how can we bring up our children?

The point that Yorm Bopha is making is that every woman needs somewhere either to nurture new life or rest their deceased bodies. Forced eviction, however, is a form of gender-based violence that reduces and unravels women's capacity to fulfil social reproductive roles that many women value. Although scholastic work collectively evidences the gendered dimensions of displacement (see Brickell and Speer 2020 for a review), *Home SOS* strengthens the case for viewing forced eviction more specifically, and emphatically, as a form of gender-based violence like domestic violence. The book reveals women's feminised responsibility to deal with the far-reaching and adverse consequences of forced eviction, from possible destruction of the family home, splintering of familial and/or community ties, the loss of livelihood, and access to essential facilities and services. These carry with them significant physical and mental burdens. As UN-HABITAT (2011a, pp. 3–4) notes and *Home SOS* demonstrates, 'the prospect of being forcibly evicted can be so terrifying that it is not uncommon for people to risk their lives in an attempt to resist or, even more extreme, to take their own lives when it becomes apparent that the eviction cannot be prevented'.

On these grounds, it is largely women who have been at the forefront of contestations over displacement globally (Casolo and Doshi 2013) and this resistance is part of the survival-work that women are undertaking. While displacement challenges the displaced 'to take their proper place [of non-being] instead of taking place' (Butler and Athanasiou 2013, p. 20), 'emplacement' (Roy 2017) as a tactic is one way that women contest the removal of their homes. The book shows, however, how these emplacement efforts are mired in gender-based violence, including state-sponsored violence against women, domestic violence, and women's incarceration as part-and-parcel of their fight for home. Direct and indirect violences are therefore inflicted physically and mentally upon women disproportionately, and this includes other deprivations of liberty that fit the definitional contours of gender-based violence. That forced eviction has been excluded from the gender-based violence radar is highly political and likely to be because it raises the spectre of complicit governments and a development model driven by capitalist dynamics and state accumulation tied to the seizing of homes at any human cost.

Indeed, seminal feminist scholarship conjectures that an indelible link exists between 'the global extension of capitalist relations and the escalation of violence against women, as the punishment against their resistance to the appropriation of their bodies and labour' (Federici 2014, p. xi). Prompted by Silvia Federici's work, Sutapa Chattopadhyay (2018, p. 1296) asks, 'how can women's vulnerability to gender-based violence be explained through numerous co-constitutive forms of violence and inequalities that shape the bodies of the marginalized?' *Home SOS* places the experiences of domestic violence and forced eviction centrefold in order to examine women's injuries, but also their survival practices, in defence of home. It therefore works to connect rather than isolate different forms of gender-based violence in its co-constitutive analysis of violences and inequalities shaping women's lives.

Converging Research Trajectories

Home SOS brings together original data from four studies I have undertaken. Over the 15-year collection period, my work has shifted from the broad-based focus of my doctorate (Brickell 2007) on household gender relations and marriage to a more concentrated engagement on domestic violence and forced eviction issues. This work began in Siem Reap Province and has since extended to Pursat Province and Phnom Penh, the country's capital.

The fieldwork for my PhD (2004–2005) was funded by the Economic and Social Research Council (ESRC) and was undertaken in the urban centre of tourist-oriented Siem Reap (Slorkram commune), home to the global heritage and tourist site of Angkor. Here female participants worked in low-paying service sector positions or were engaged in the home-based selling of soup and refreshments. The other fieldsite was located in its rural vicinity where rice farming still predominated (Krobei Riel commune) and where women were mainly subsistence farmers and weavers.[4] The research was based on 100 oral histories and 40 semi-structured interviews. The sample of interviewees was based on an equal proportion of men and women of differing ages living in the two communes. Participants were sought through visits on different days and times to randomly selected individual households in the two communes. Unlike the semi-structured interviews, the oral histories were completed in two sessions (lasting between two to four hours each) and were used to understand transitions over the life course given the shifting expectations, responsibilities, and attitudes tied to the roles of men and women in and outside the household. These were ordinarily carried out in participants' homes out of the sun and, given their length, women sometimes undertook piecework or housework. A profile was also used to gather background socioeconomic information about respondents and their respective households. In total, during my doctoral research, I interviewed 19 women in Slorkram and 5 women in Krobei Riel who had directly experienced marital breakdown, the majority of whom had encountered domestic violence in their spousal relationship(s). This included Orm, whose life story had a profound influence on the development of my research trajectory and this book.

My first encounter with Orm came across the metallic countertop of her sugar cane juice stall positioned outside her house in Slorkram Commune, Siem Reap (Figure 1.2). She had lived there with her two young daughters for

Figure 1.2   Orm's sugar cane juice stall outside her home, Siem Reap, 2004. Photo: Katherine Brickell.

4 years. Sipping the sweet liquid she had just squeezed through her press, we went on to spend several hours together talking. The joyous shrieks of the boys jumping in the river behind us felt in stark relief to the sombre tone of her voice as it began to crack in this initial interview. Surrounded by photographs of her standing alone, Orm quickly started telling me about the social discrimination she encountered as a divorcee, and the extreme depression that resulted. This extended to anxiety-induced weight loss and the attritional wearing down of her body in response.[5] Her mood was bleak and her failed marriages inflected with sadness and worry that had become an affective presence in her life.

Orm's interview revealed the reluctance to talk publicly about her circumstances on the basis of perceived guilt, which upholds the idea that women should not take fire outside the house, but let it smoulder inside. Crises should, by this prevailing logic, be contained as internal domestic affairs and not leaked into the public sphere. What was also telling for me in Orm's interview was how the failings of her intimate relationships and her insecure tenure status residing on the riverbank without legal title (Figure 1.3) compounded her overarching sense of precarity and outsider status. Much like the language used to describe the history

Figure 1.3   Exterior of Orm's home, Siem Reap, 2004. Photo: Katherine Brickell.

and circumstances of her intimate relationships, she explained that she had no choice but to 'bear living here'. Orm was deeply unhappy because she did not know if and when the land would be confiscated and frequently noted her constant fear of her eviction 'being soon'. Earning only 40 US dollars per month and living alone without a partner, she noted that there was little way of turning her life chances around. This slow and anticipatory emergency was never far from her mind. Crisis had an omni-presence in her life history that she keenly detected.

Orm's interview impressed upon me the toll that the responsibility of managing the crisis ordinary people can have, and how this weight is linked memories and experiences of the past. When fantasies of 'the good life' become untenable (Berlant 2011), as they had for Orm, a sense of crisis can pervade to such an extent that life feels hopeless. Orm describes that she 'can only hold the pain in my chest' and feels that 'trouble' was 'accumulating continuously' in her life. This inability to move forward positively in life, the sheer relentlessness of issues amassing, and the quotidian internalisation of this harm, is etched into the crisis ordinary. To escape it, suicide is one route Orm contemplated. The only thing deterring her suicide was concern for her children's well-being. Orm had lost both her parents when she was a child during the Khmer Rouge regime and after being forcibly displaced to Battambang Province lived with her grandmother before she was also killed. Moved regularly across the country, Orm variously laboured on rice fields, broke up termite mounds, and built irrigation channels, all under the regime's orders. She did not wish her daughters to be orphaned as she had. In her first marriage, she explained that her husband's parents did not accept her because she was an orphan. Often, she said, looking down, 'they didn't allow me to sleep inside with my husband, closing the door on me' and forcing her to sleep outside. Orm's desire for home and acceptance had once again been stymied.

For Orm, the ensuing crisis was not only an event but felt like 'an enduring condition of life' (Roitman 2013, p. 2) that she could not escape however hard she tried. This is in large part because of the normative ideals and constraints to which she felt held as a Cambodian woman who should be happily married and living in a stable conjugal home. In Cambodia these seemed available only to certain populations she was stigmatised by, and excluded from, on moral and financial grounds. That crisis 'entails reference to a norm' (Roitman 2013, p. 4) was especially pertinent to Orm's gendered story of the crisis ordinary.

My doctoral research provided the basis for a project four years later on marital dissolution, an issue of intimate significance to Orm and several other women I had met during my PhD research. Supported by a Royal Geographical Society small grant, I returned to the rural Krobei Riel commune in 2011. There I carried out a total of 22 in-depth interviews with ever-abandoned, separated, or divorced women.[6] During the month-long research, all women from the commune who were identified by village-level leaders as falling into the aforementioned categories were interviewed through this targeted sample approach. Working across this range of statuses was important in a country where marriage and its end can take multiple forms. The 2005 Cambodian Demographic Health Survey (National Institute of Public Health et al. 2006, p. 266) suggests that only one-third of ever-married women (652) canvased (2162) had signed a contract in front of commune authorities.[7] Research by Robin Biddulph (2011) on what are termed 'de facto marriages' suggests that whilst couples apply for permission from the commune and follow formal public announcement guidelines, they do not complete the registration process at the commune to gain marriage certification. Whilst not legally verifiable, these marriages are based on community recognition and also on family books compiled and issued by local authorities. As Mehrvar, Kim Sore, and Sambath (2008, pp. 14–15) have stated in their work on land rights in Cambodia, 'Most women interviewed did not understand the difference between divorce in a legally registered marriage and an agreement between parties to dissolve traditional marriage (separation).' Consequently, despite a decision from the court being required for a 'formal divorce' in cases of legally registered marriages, they found official court decisions for divorce to be extremely rare.

Given this complex legal picture of marital dissolution, a profile form was again used to collect data on marital status, age, household structure pre- and post-breakdown(s) (including intermediary arrangements), and engagement in paid work (again with notes to any changes). The women ranged in age between 24 and 48 years old. Whilst the project predominantly focused on Krobei Riel, I also spent time attempting (largely in vain) to relocate the 19 women who had experienced marital breakdown identified in my doctoral research in the Slorkram commune. I did, however, manage to successfully recontact and re-interview Orm with the help of a photograph I had kept from seven years prior.

Orm had moved further down river to allow for what she was briefed as bridge widening works. Despite Orm's tumultuous relationship history, out of stigma and shame, mixed with a desire for emotional and financial support, Orm had also recently remarried. Yet this change in her marital status had not removed the need for survival-work to support her family and she again felt 'as usual' depended *upon* rather than being depended *for*. Describing being 'looked down' upon and 'scorned', her decision to remarry was in part influenced by her 'out of place' status. Orm's reluctant choices emerged from her attempts to conform to acceptable Khmer womanhood. In this sense, symbolic dimensions of marital dissolution – in which discrimination plays a part – are shown to directly shape decision-making processes. The 'choice' to remarry becomes a tool to regain status, acceptance, and self-confidence at the same time as familial stability and happiness.[8] Prevailing gender norms are important factors to understanding the 'home unmaking' (Baxter and Brickell 2014) and remaking practices of Cambodian women who have experienced marital breakdown (Brickell 2011). They are in one sense contradictory, limiting the agency of women, to challenge the spectre of violence they encounter in cases of domestic violence and forced eviction. Yet, as Orm's marital manoeuvres testify, they can provide a form of moral legitimacy and protection that women mobilise in situations of discrimination.

A year later after I conducted this research, and having re-met Orm, I began leading a study (2012–2015) joint funded by the ESRC and Department for International Development (DFID) entitled 'Lay and Institutional Knowledges of Domestic Violence Law: Towards Active Citizenship in Rural and Urban Cambodia'. It explored the (in)efficacies of the country's first ever DV law, 'The Law on the Prevention of Domestic Violence and the Protection of Victims' (Royal Government of Cambodia (RGC) 2005) (see Figure 1.4). Each research element was conducted

Figure 1.4   Domestic violence law poster in Pursat Province, 2013. Photo: Katherine Brickell.[9]

in eight communes divided equally between urban and rural environs of two provinces – Pursat and Siem Reap (including Slorkram commune in which I had undertaken my prior research). In early 2013, the qualitative research was conducted that forms the backbone of the domestic violence (DV) data I engage with in the book. It included 40 interviews with female DV victims (identified by NGOs and community authorities), 40 with an equal proportion of male and female householders (approached randomly), and a final 40 with a range of different stakeholders who had an occupational investment in DV reduction. This latter sample included legal professionals, NGO workers, police officers, and other authority and religious leaders. The viewpoints of such stakeholders are a distinctive and necessary feature of the empirical chapters that follow. They help 'to reveal specific features of contentious politics in the terrain of law, since as an "institutional environment" law makes mediators and translators crucial to the activation of a politics of rights' (Woodman 2011, p. 190). Their professional behaviour and conduct also sends out powerful wider messages to victims, offenders, and the wider society about the prospects of women gaining justice (Briones-Vozmodiano et al 2014; Lila, Gracia, and Gracia 2013). In January 2019 I undertook follow-up meetings with key stakeholder organisations in Phnom Penh.

These three projects featured in *Home SOS* are joined by a fourth, on forced eviction. The need for this research became more pressing over time as I undertook the other studies. Forced eviction was a human rights violation, which was becoming ever more present across Cambodia. It was also something that Orm spoke to me about previously in the 2011 study on marital dissolution in Siem Reap. In 2013, I returned to Siem Reap to re-connect with Orm once again. I could not find her, however, and her riverside home had disappeared. The year prior, nearly 400 families had been evicted from their homes in the Slorkram commune of the city. Authorities had justified the evictions to develop, further widen the river, and make new communal gardens (Transterra Media 2012). Despite being told that they had until 1 April 2012 to move, on 27 March a large police force arrived in the early morning and demanded families and businesses leave immediately (Figure 1.5).

Figure 1.5   Slorkram river community before (26 March) and after (27 March) the eviction, Siem Reap, 2012. Source: Courtesy of Philippe Ceulen.

Orm's home had been displaced several times; its fixity to the river bank had gradually been unstitched until it was gone. The fire that tore through it was deliberately ignited to remove every trace of familial life once lived on Siem Reap's river banks. It was the end point of an attritional war waged by provincial authorities to make room for more 'profitable' uses. Described as a 'sore sight for tourist eyes', the river 'clean up' (Sokha 2006) was justified to help beautify the city, the economy of which is driven by international tourists visiting the archaeological ruins of Angkor close by.

The spectre of forced eviction and the eradicative violence that accompanied it, also revealed fault lines that were hard to ignore and compelled the twin study of domestic violence and forced eviction. Their juxtaposition was something I kept on returning to in my thinking. While the passing of DV law suggested a political willingness, of sorts, to tackle this type of violence against women through 'rule of law', the Cambodian government were concurrently using 'rule by law' against women contesting forced eviction on the streets of the country's capital (see Chapter 6 for a discussion of these distinctions). 2011 saw the escalation of violences in, and politically motivated charges against, the Boeung Kak Lake (BKL) community of Phnom Penh and women in particular. Law and violence had an intimate relationship in Cambodia and was one I felt needed exploring. Over the years, spending more time in Phnom Penh for the DV law study interviewing NGOs and policymakers, I took the opportunity to visit BKL in 2012 and then to start new work there. This fieldwork was carried out in 2013 and 2014.

Garnering both national and international attention, BKL is perhaps the most widely known case of forced eviction, and collective resistance to it, in Cambodia. The controversial project involved the eviction of families living around the lake over many years in central Phnom Penh (see Chapter 4 for more information; Nam 2011 on its history and Kent 2016 on recent events there). As the United Nations General Assembly (2012, p. 8) note, 'The case is emblematic of the desperation that communities throughout Cambodia feel in resolving their land disputes, and the ensuing civil unrest.' To understand forced eviction as an embodied, located, and grounded phenomenon, in-depth interviews were conducted with 15 women from BKL who had either been forcibly evicted and/or who had become active participants in protest in 2013. In addition to these interviews, material was gained via audio-recorded tours (Figure 1.6) with some of the women as well as more informal 'hanging out' at the women's advocacy centre. All the women chose to be identified using their real names: Kong Chantha, Bo Chhorvy, Phan Chhunreth, Nget Khun, Ngoun Kimlang, Soy Kolap, Srei Leap, Heng Mom, Van My, Sear Naret, Srei Pov, Soung Samai, Phorn Sophea, Srey Touch, and Tep Vanny. Given the government's intensified clamping down on freedom of speech and willingness to incarcerate BKL women in recent years since, I have not used their specific names where I refer to any direct criticism of government party members.

Figure 1.6   Audio-taped tour of Boeung Kak by residents, 2013. Photo: Katherine Brickell.

A further five interviews in 2013 were conducted with women living in Trapaing Anhchanh, a resettlement community made up of evicted residents from the 'Cambodia Railway Rehabilitation Project'.[10] These are featured in Chapter 4 using pseudonyms. In 2014 I returned to interview three BKL women, Tep Vanny and Srei Leap, who I had met previously, and Yorm Bopha, who was in prison during an earlier research trip (see Chapter 6 for more information on incarceration). I also extended my sample to include six husbands of BKL activists and other menfolk to understand their experiences (I again adopt pseudonyms in the empirical chapters). Across both years of research, I undertook six further interviews with NGOs involved in land rights issues. In 2017 and into 2018 I also met several Cambodian political figures and journalists who were in London and who offered their views on the BKL situation.

Overarching Methodological Approaches

Taken together, the four projects I have outlined are marked by a series of overarching methodological approaches. First, the material presented in the book brings together and scans across a 15-year horizon that is complementary to feminist geography research and emphasises the benefits of sustained engagement, of long-term relationships, and commitments to particular groups of people or issues (McDowell 2001; Valentine 2005). It is not only that my arguments are based on a larger volume of interviews than individual studies might yield (over 300 in total) but that their undertaking over more than a decade gives rise to insights that go beyond snapshots of time. As Clare Madge et al. (1997, p. 96) note, longitudinal studies are valuable 'because the research is not fixed in time but reflects the dynamism of people's experiences, so enabling temporal differences' to be comprehended (see also Sou and Webber 2019). The influence of Orm's life story, followed from 2004 until 2013 inspired, for example, the shape of my research in the years that followed in this book. Learning from her twin experiences of marital breakdown and forced eviction, cemented my thinking that rather than discuss domestic violence and forced eviction in isolation from each other, it was important to trace and understand their connective tissue. Being attentive to individual as well as collective trajectories in a country in flux offers the ability to explore how women's lives and those of their families are swept up with Cambodia's transition – 'to tell a story capable of engaging and countering the violence of abstraction' (Hartman cited in Saunders 2008, p. 7). Inspired by Orm, the book is therefore less reactive to, and more reflective on, the (un)eventful and women's survival-work on a daily and long-term basis. The book's ability to capture the slow violences, less perceptible traces, and repeated articulations of injustice in Cambodian society are heightened through this approach. They provide a fuller and more cumulative story about the violences encountered in domestic life than might otherwise have been possible. Furthermore, given that violence is 'a processual and unfolding moment, rather than an "act" or "outcome"' (Springer and Le Billon 2016, p. 2), the longitudinal allows the book to transcend the singular event and moment to demonstrate the processing and unfolding nature of violence.

Second, the book draws out the experiences of ever-married Khmer women from the wider data set of participants possible from the four studies.[11] Albeit selective, this approach aims to ensure that histories of women in Cambodia are told not only through the more formal realms of female politicians (Lilja 2008) or Buddhist nuns (Kent 2011a, 2011b) but also through more informal spheres of political power tied to the marital realm. The book's focus on the Cambodian housewife as domestic violence survivor and grassroots activist contributes to work on the history of women and power in the country (Jacobsen 2008). It works to ensure that as domestic figures they are not written out of history either as unexceptional or rebel figures who are trivialised and/or derided in the popular or scholarly imagination. My pivot around marriage and its breakdown also arises from what is described as 'the near universality of marriage in Cambodian society' (National Institute of Statistics (NIS) et al. 2014, p. 98). Reflecting the wider centrality of marriage in Asia (Piper and Lee 2016), figures record that the proportion of women in Cambodia who have never married decreases with age to a low of only 4% among those aged 45–49 (NIS et al. 2014, p. 98). While this ubiquity does not mean that I exclude other perspectives and voices from the data in the book, it does mean that *Home SOS* foregrounds the violences visited upon, and told through, the narratives of middle-aged women who have been, at one point, in a spousal relationship.

Third, each of the projects outlined has been driven by a commitment to in-depth qualitative research, namely interviews, to better understand 'geographies that wound' (Philo 2005) both in terms of their affliction and treatment. In the face of what Russell Hitchings (2012) has identified in human geography as a growing hesitance about interviews to research everyday life, I reaffirm the vital and enduring role of 'talk' for learning about domestic life and its ruptures. This falls in line with feminist geographical research which has a long history of engagement with interviews (Longhurst 2016). Given potential suppression and eradication of victims' capacity for speech within and beyond the Cambodian family, interviews encourage the narration of stories to communicate their experiences. These recorded Khmer language interviews were transcribed and then translated verbatim into English by research assistants I worked closely with in each project. In my doctoral research, I had a basic command of Khmer, and worked with two male interpreters who were students at the time, and who supported my broad-brush explorations of household life. Female research assistants undertook the marriage dissolution and forced eviction research with me. Meanwhile, following World Health Organization (WHO) guidelines on the ethical conduct of research on domestic violence, the interview team in the larger DV law study included one male and one female research assistant who carried out the research with male and female respondents (respectively) after a period of training, piloting, and initial in-the-field supervision. At the start of each interview informed-consent was recorded on the digital voice recorder, and with my more sensitive postdoctoral research, it was explained to respondents that if at any point they felt distressed and wanted to change the subject (or stop the interview), then this was acceptable. Information on NGO assistance was additionally given to two participants who requested help in the DV law study.

Structure of the Book

In the next chapter I continue to set up the conceptual parameters through which the interconnected fires of domestic violence and forced eviction can be best understood. Chapter 2 brings into dialogue scholarship on the crisis ordinary with bio-necropolitics and precarity, intimate war and slow violence, law and lawfare, and rights to dwell. It therefore harnesses and aims to advance a diverse yet synergistic set of theories from geography and cognate disciplines to impress the political significance of threatened homes, women's survival-work, and the paramountcy of gender-based violence.

Chapter 3 provides information on Cambodia's turbulent history and some of the key social-spatial continuities and junctures of change being witnessed in the country today. The chapter thereby grounds the book in the political economy of Cambodia. As feminist scholars argue, political economy matters; it 'highlights the patterns of material power and relationships that profoundly affect both the prevalence of violence and insecurity and the efforts to eliminate them' (True 2012, p. 7; Elias and Roberts 2018). Just as violence within the family 'is a multifaceted phenomenon grounded in the interplay among personal, situational, and social cultural factors' (Heise 1998, p. 262), so too are the paths of domestic violence and forced eviction interventions embedded in, and mediated by, an environment of interconnected factors (Adelman 2004; Brickell 2015; Fulu and Miedema 2015; Sjoberg 2015; True 2012). As I advance through the book, domestic violence and forced eviction are present and felt in the home precisely because of their rooting in other sites and scales of power (Pain and Staeheli 2014). Detailing some of the 'political and institutional context in which the *production* of precarity occurs' (Waite 2009, p. 421, emphasis in original) is therefore core to the aims of Chapter 3.

Chapters 4, 5, and 6 present the empirical material collected. In each I draw out commonalities and differences between the rights violations of domestic violence and forced eviction. My aim is to provide the reader with an ongoing and at times heightened sense of tension, dissonance, irony, and anger as the two unfold against each other in respect to their SOS calls (Chapter 4), responses (Chapter 5), and outcomes (Chapter 6). These are long and at times overwhelming chapters, which are shaped by the gravity of the subject matter and the weight of empirical material from both domestic violence and forced eviction victims on their experiences.

More specifically, Chapter 4 pursues a bio-necropolitical reading of contemporary Cambodia to explore women's narratives of domestic violence and forced eviction. Their narratives attest to truncated homes and lives subject to political abjection. I explore how these domestic crises are intimate wars that Cambodian women traverse through diverse encounters with death: from suicidal contemplation and suicide; avoiding and/or dwelling in social death; feeling like the living dead; to enduring feticide (the destruction of a foetus). The chapter is attentive to the resilience shown by women (and a number of featured men) in situations of domestic violence and forced eviction and how this survival-work is formative of the crisis ordinary.

Chapter 5 critically examines the actors and mechanisms through which the violences detailed in Chapter 4 have been responded to. While in the case of domestic violence I centre on the official state-sanctioned recourse to new legislation, my attention to forced eviction pivots on unsanctioned and creative routes to official redress via women's grassroots activism. I explore both these refusals for 'fire' to be contained to the home and study the extra-domestic spaces of parliament, the courts, and the street to pursue this. I thereby focus on the (un) invited and (un)eventful spaces of resistance of citizenship that have emerged from the home as an expansive and multi-dimensional space of political and legal significance. The chapter's focus on the different meanings given, commitments made, and actions taken to address domestic violence and forced eviction forms the groundwork for Chapter 6, which analyses their outcomes for women's lives.

Chapter 6 shows how women's experiences of violence can galvanise resistance but that this can also heighten their exposure to further violence and jeopardise their survival. Here I demonstrate the significance of law and lawfare for understanding the consequences of the survival-work explored in the two chapters preceding it.

Chapter 7 provides the conclusion to the book and summates how the home and the women they notionally shelter are imbued with variable levels and logics of disposability and grieveability by the Cambodian government and society in general. The myriad violences that women are encountering, I conclude, are not accidental to the governance of contemporary Cambodia but are central to its functioning. Domestic violence and forced eviction are, therefore, not unrelated typologies of gender-based violence. They are interrelated oppressions and brutalisations of domestic life that gravitate and retrain multiple subfields of geography on to the home sphere as a public–private hybrid worthy of energised study.

Endnotes

1 Ever-married refers to persons who have been married at least once in their lives although their current marital status may not be married.

2 Linguistically, at least in Khmer, there is no word for home. While *phteah* literally means house, the Khmer word *Kruosar* is equivalent to family.

3 These high-level observations relate to broader academic ones that the effects of displacement are not gender neutral, impacting men and women differently. This has been evidenced in relation to: conflict (Brun, 2000; Ensor 2017; Majidi and Hennion 2014; Manchanda 2004; Meertens and Segura-Escobar 1996; Suerbaum 2018); development (Brickell 2014a; Fernández Arrigoitia 2017; Mehta, 2009; Muñoz 2017, Perry 2013); gentrification (Lyons et al 2017; Mirabal 2009; Maharawal and McElroy, 2018; Sakizhoglu 2018; Watt 2018; Wright 2013); the extractive industries (Ahmad and Kuntala Lahiri-Dutt 2006); and/or natural disasters (de Mel 2017; Gorman-Murray, McKinnon, and Dominey-Howes 2014; Juran 2012; Tanyag 2018).

4 More detailed information on each vicinity can be found in my PhD thesis (Brickell 2007).

5 At this point in time, Orm had one prior formal divorce and two subsequent failed non-marital relationships.

6 The original research design had intended to include an equal proportion of men in the sample. However, no village leaders were able (or perhaps willing) to identify men who had experienced marital breakdown. National level statistics available at the time corroborated this situation with more women officially 'divorced' than men in both the 2008 General Population Census (0.8% males/3.1% females) (National Institute of Statistics 2009) and the 2005 Cambodian Demography Health Survey (CDHS) (1.8% males/4.2% females) (National Institute of Public Health et al. 2006). More recently, the National Institute of Statistics et al. (2014) record that, on average, across the 15–49 age cohort, only 1.4% of men are divorced in comparison to 3.4% of women.

7 This information is not available in newer versions of the CDHS.

8 See, however, the study by Sothy Eng, Szmodis, and Grace (2017) on the prevalence of domestic violence among remarried women in Cambodia.

9 The translation of this poster, is as follows. (Left): The abilities and duties of the authorities: (in the case of) the use of violence against wives or family members, the closest authorities have a duty to intervene immediately in cases where violence against the family takes place. (Right): Duties of the authorities when intervening in family violence: the authorities must identify a clear reason for the occurrence and then report it urgently to the prosecuting authorities (the court).

10 The Asian Development Bank and AusAid have attracted criticism for inadequate safeguarding against breaches of human rights as part of their funding of the 'Cambodia Railway Rehabilitation Project'. The range of options and full entitlements under the resettlement plan were not made available to project-affected households (Inclusive Development International 2012). The project was launched in 2006 to restore the country's dysfunctional railway infrastructure as part of the ADB's Greater Mekong Sub-Region Programme. See http://www.babcambodia.org/railways for a video on the harm experienced.

11 I refer to 'Khmer' women specifically as I did not conduct research within any ethnic minority communities who represent less than 5% of the population.

Chapter Two
Conceptualising Domestic Crises

Introduction

In Chapter 1, I introduced the overarching ways that the book synthesises its analysis of domestic violence and forced eviction, as crisis ordinaries, as sources and forms of survival-work, and as types of gender-based violence, which have been unduly overlooked in academic writing. Over the next five sections, the crisis ordinary and survival-work, bio-necropolitics and precarity, intimate war and slow violence, law and lawfare, and rights to dwell, I continue to set up the conceptual parameters through which domestic violence and forced eviction can be best understood. Their exchange and prognosis-seeking contribute to efforts in geography to trace and explain the intimate violences of world (un)making and their multi-origin and multi-scalar manifestations.

Crisis Ordinary and Survival-Work

While Rebecca Solnett (2009, p. 10) writes that 'emergency is a separation from the familiar, a sudden emergence into a new atmosphere, one that often demands we rise ourselves to the occasion', the crisis ordinarily frames domestic violence and forced evictions as long emergencies of slow violence that are unrelenting in their normative unfolding. They are domains 'where an upsetting of living is revealed to be interwoven with everyday life after all' (Berlant 2007, p. 761). Eminent

Home SOS: Gender, Violence, and Survival in Crisis Ordinary Cambodia, First Edition. Katherine Brickell.
© 2020 Royal Geographical Society (with the Institute of British Geographers). Published 2020 by John Wiley & Sons Ltd.

to Cambodia, Naly Pilorge, Director of the Cambodian NGO LICADHO, discerned more than a decade ago that 'everyone claims Cambodia has come through a period of barbarism, but the sadism is still bubbling beneath the surface. Extreme violence, greed, and disregard for the most basic rights – of giving people a place to live – are still with us daily' (cited in Levy and Scott-Clark 2008, np). *Home SOS* thereby explores the paths of Cambodian homes and lives (still) submerged in protracted circumstances of violence. These are paths that in various ways have global reach and wider relevance beyond Cambodia.

Domestic violence has been described as 'one of the starkest collective failures of the international community' (Action Aid 2010, p. 1). It is likely that there has been an under-estimate, given the non-reporting of cases, of worldwide data, which indicates that 35% of women have experienced either physical and/or sexual intimate partner violence or non-partner sexual violence (World Health Organization 2014a). In the last 40 years, violence against women has become visible as a major social issue and has been labelled 'a global health problem of epidemic proportions' (WHO 2014a). Domestic violence has also been described in the media and international health arena as a 'hidden crisis' worldwide (BBC 2018; WHO 2014b), thus rhetorically challenging its non-eventful status in popular and political consciousness. The 'hidden crisis' of violence against women has additionally been raised within the context of other, more archetypal crisis scenarios, including academic work on the current refugee 'crisis' in Europe (Freedman 2016) and in the wake of natural disasters (Nguyen 2019; Parkinson and Zara 2013).

In a similar discursive vein to domestic violence as a global problem, forced eviction – when people are forced out of their homes against their will often with the threat of use of violence (Amnesty International 2012, p. 2) – has been described as a 'global phenomenon' and a 'global crisis' (UN-HABITAT 2011b, p. viii). Forced eviction is 'the permanent or temporary removal against their will of individuals, families and/or communities from the homes and/or land which they occupy, without the provision of, and access to, appropriate forms of legal or other protection' (Committee on Economic, Social and Cultural Rights 1997, np). Every year, millions of people around the world are forcibly evicted from their homes and their land (United Nations 2014). Yet, despite this, for too long 'social scientists, journalists, and policymakers all but ignored eviction, making it one of the least studied processes affecting the lives of poor families' (Desmond 2016, pp. 265–296). Looking to rectify this trend, Matthew Desmond's (2016) book *Evicted* shows that in Milwaukee, the United States, eviction from rental accommodation has become commonplace for women in its poorest black neighbourhoods; and, while 'incarceration has come to define the lives of men from impoverished black neighborhoods, eviction was shaping the lives of women. Poor men were locked up. Poor women were locked out' (p. 98). Although the young and the old, the sick and the able-bodied, are not unaffected by eviction, he contends that for the women in his ethnographic research, eviction had become 'ordinary'. 'Walk into just about any urban housing court in America', Desmond observes, 'and you can see them waiting on benches for their case to be called' (p. 299).

Research on domestic violence and forced eviction reveals the structural conditions and power geometries that render these violences chronic features of women's everyday experiences across the globe. These all-too-familiar violences can be considered 'as' crises, and reflect home life 'in' crisis (see Roitman 2013, p. 2 on distinctions of crisis). Yet domestic violence and forced eviction are not exceptional events; rather, they are emblematic of pervasive precarities and displacements lived in and through the domestic sphere. On balance, they do not attract 'feverish crisis pronouncements' or reach the heights of 'clamorous crisis' like other more visible and visibilised crises (Roitman 2013, p. 6). Instead, they are propelled by 'longitudinal forces of upheaval' (Ramsay 2019a, p. 4), which calls upon women's survival-work in a grueling intimate war mired in patriarchal and violence social, economic, and political relations. Their scale and scope, their diffusion of trauma into domestic life, have become 'increasingly normal and perpetual instead of functioning as localized disruptions to the ordinary' (Calvente and Smicker 2017, p. 3). As Ayona Datta (2016a, p. 329) writes in relation to Delhi slums, 'violence is constructed not as an interruption of intimacy but rather as a route through which intimate relationships are upheld, sustained and rendered ordinary'.

In a connected sense, *Home SOS* positions domestic violence and forced eviction as forms of structural violence because they are embedded in a political economy of inequality and violence that causes injury and unnecessary death. Johan Galtung (1996, p. iii) describes structural violence as 'the violence frozen into structures, and the culture that legitimizes violence'.[1] This freezing is formative of the crisis ordinary but is not summative of it. The remit of the crisis ordinary is not limited to structural violence as a normalised condition. Rather it encompasses the management and endurance of rupture in everyday practices and performances of domesticity. It is here that an enlivened and embodied sense of structural violence emerges. As Achille Mbembé and Janet Roitman (1995, p. 325) write,

> … it is in everyday life that the crisis as a limitless experience and a field of the dramatization of particular forms of subjectivity is authored, receives its translations, is institutionalized, loses its exceptional character and in the end, as a 'normal,' ordinary and banal phenomenon, becomes an imperative to consciousness.

While a swathe of recent publications on crisis point to the significance of 'intimate uncertainties' in reproduction (Strasser and Piart 2018, p. v), the crisis ordinary underscores, in contrast, the predictability and embeddedness of these uncertainties. As Berlant (2007, p. 760, emphasis in original) notes, when scholars and activists refer to 'long-term conditions of privation, they choose to misrepresent the duration and scale of the situation by calling a *crisis* that which is a fact of life and has been a defining fact of life for a given population that lives in that crisis in ordinary time'. Crisis, in this vein, has 'come to be construed as a protracted historical and experiential condition', an 'enduring crisis' that questions its designation as a 'critical, decisive moment' (Roitman 2013, p. 2). That crisis 'has become part of the infrastructure of the ordinary' (De Abreu 2018, p. 747) runs the risk, however, of being normalised rather than questioned within an ontological condition of uncertainty. Roitman (2013, pp. 95–96) is right to query that 'If history amounts to a record of interruptions (suffering, alienation, crisis) how does one successfully resist or avoid the temptation to achieve admission into the record, thus severing recognition and noteworthiness from the achievement of politics?' The crisis ordinary, however, is both a marker of crisis and of the ordinary; it encourages scholars to connect and contest subterranean structures of violence with those lived closer to the surface in the extra-domestic home. The survival-work populating the pages of *Home SOS* responds not only to crisis in the ordinary, but also actively extends it to think more about the crisis of the ordinary and its contested reproduction.

Home SOS focuses accordingly on how women get by with, pragmatically adjust to, and/or confront violence in different times and spaces and using different practices and consciousness of survival. Although 'under a regime of crisis ordinariness, life feels truncated – more like doggy paddling than swimming out into a magnificent horizon' (Berlant 2007, p. 779), the book takes a more variegated approach to agency than this analogy perhaps communicates. A fuller understanding of women's survival-work mobilises distinctions between resilience, reworking, and resistance (Katz 2001). Katz's (2001) contextualized accounts of agency differentiate between 'resistance' (oppositional consciousness that achieves transformative change), 'reworking' (that alters the organisation but not the polarisation of power relations), and 'resilience' (that allows people to survive but with limited change in circumstances) (see also Chapter 5). Although Berlant's notion of doggy paddling evokes a sense of resilience above all else, what the crisis ordinary does, for me at least, is to mark out and critically, *call out*, the diversity of survival-work, be this resilience, reworking, and/or resistance, as a crisis in itself.

As such, the crisis ordinary sidesteps the resounding attraction that resilience discourse, in particular, seems to hold for many governments, policymakers, and practitioners. Resilience is 'tightly bound to the adage that we now live in a "time of crisis"' and has come to 'stand for the ability to absorb, withstand, persist and even thrive and reorganize in the face of the shocks and disturbances of always uncertain becoming' (Simon and Randalls 2016, pp. 3–4). The zeitgeist of resilience risks normalising the status quo, however, and having a depoliticising effect by retaining and potentially deepening long-noted expectations on women to behave as resilient subjects consolidating the function and structure of capitalist patriarchy. That women from the Global South are constructed as resilient subjects in policymaking who can manage the responsibility of bringing home money and providing care and services for family members is also widely noted and pertinent here (see, for example, Erman and Hatiboğlu 2017). For example, in the context of Typhoon Haiyan in the Philippines, which left a devasting wake in 2013, Maria Tanyag (2018) argues that disaster resilience measures have further divested responsibility for disaster response from the state to the household and community levels. In so doing, she argues that the assumed elasticity of women's unpaid labour and the propensity for self-sacrificing behaviour 'has served to reinforce the structural roots of gendered vulnerability' (Tanyag 2018, p. 566). Tanyag explores this vulnerability through mobilising the concept of Shirin Rai, Hoskyns, and Thomas (2014, pp. 88–89) of 'depletion for social reproduction', when 'resource outflows exceed resource inflows in carrying out social reproductive work over a threshold of sustainability, making it harmful for those engaged in this undervalued work'. This depletion 'is core to what is understood as a crisis of social reproduction, that is, the inability of people to adequately reproduce their livelihoods' (Dowling 2016, p. 455) and homes.

Whilst emerging feminist work on depletion (Gunawardana 2016; Dowling 2016; Fernandez 2018; Rai et al. 2014; Tanyag 2018) does not reference Berlant's thinking in its theoretical articulation, there are complementarities that can be productively made with her work on the crisis ordinary. Writing on 'slow death', for example, Berlant (2007, p. 734) contends that the physical wearing out of a population takes place 'when that experience is simultaneously at an extreme and in a zone of ordinariness, where life building and the attrition of human life are indistinguishable'. Women's disproportionate burden for mitigating violence and crisis has, therefore, a likely depletive effect. The very undertaking of the physical and emotional survival-work necessary in circumstances of domestic violence and forced eviction, becomes normalised in their home making and life building efforts. Revealing the gendered geographies of death and the survival-work of living on are therefore rendered more possible by being attentive to the understudied spatialities and temporalities of the extra-domestic and its crisis ordinariness. The long-term crisis-management work of domestic violence and forced eviction is therefore replete with (further) risks of bodily depletion.

On these grounds, there is unexhausted conceptual merit in thinking through intersections between survival-work and slow violence. Interdisciplinary feminist work also has an important role to play in advancing geographical literatures on crisis and emergency that are showing an accelerated pace of interest in slow violence (as recent examples, see Anderson et al. 2019; Brydolf-Horwitz 2018; Rydstrom 2019; Pain 2019). These examples emphasise how 'psychic and physical attenuation and deterioration are part of the ongoingness of ordinary life' which 'allows us to decouple the concept of slow emergencies from the concept of the event, or more precisely one particular mode of eventfulness associated with the sensational and the spectacular' (Anderson et al. 2019, p. 11). Feminist perspectives can really push, probe, and propel the theorising and addressing of masculinist practices, connected drivers, and social relations of the crisis ordinary (e.g. patriarchy and capitalism) that structure these attenuations and deteriorations from a gendered perspective. The survival-work that *is* the crisis ordinary represents an important conceptual route to register the social reproductive labour that sustains and, in certain circumstances, challenges the chronicity of domestic crises.

Bio-necropolitics and Precarity

If the crisis ordinary and survival-work draws closer attention to women's experiences of home life, then bio-necropolitics and precarity bring governmentality into ever greater view. Domestic violence and forced eviction are neglected yet essential parts of debate on bio-necropolitics. While 'biopolitical powers work to manage, order and foster life for citizens worthy of protection, such powers work in tandem with necropolitical powers that produce death for those destined to abandonment, violence and neglect' (Lamble 2013, p. 242). Taking on board recognition of biopower and necropolitics as 'two sides of the same coin' in which 'the explosion of discursive interest in the politics of life itself, in other words, affects also the geo-political dimension of death and killing' (Braidotti 2013, p. 9), the book facilitates discussion on women's experiences of home lived at the edges of life and death. It draws on, and encourages, feminist thinking on the home and female body as a privileged target for the power-techniques and power-relations of male-centred systems of exploitation (Federici 2004).

In his seminal post-colonial writing, Achille Mbembé (2003, p. 40) outlines the parameters of necropolitics and anchors it to 'the various ways in which, in our contemporary world, weapons are deployed in the interest of maximum destruction of persons and the creation of *death-worlds*, new and unique forms of social existence in which vast populations are subjected to conditions of life conferring upon them the status of *living dead*' (emphasis in original; see also Mbembé 2019).[2] According to Thom Davies, Isakjee, and Dhesi (2017, pp. 5–6), necropolitics arose as a 'reaction to the inadequacy of biopower to conceptualize the more extreme cases of body regulation, when life was not so much being governed, as much as death itself was being sanctioned'. For Michel Foucault (1997), biopolitics was foregrounded on the social body as an object of power whose life is secured, managed, regulated, and intervened in, through regimes of authority. Matters of biopolitics, therefore, intimately relate to care, marriage, and family in which human life exists inside, rather than outside, political processes. Foucault's theoretical propositions, however, have been criticised for being too rooted in discursive practice rather than social and economic relations that identify the source and motivations of power techniques to administer and promote life (Federici 2004). Foucault also had little to say about the politics of letting die, 'why governing authorities would elect not to intervene when they could, or select one subset of the population for life enhancement while abandoning another' (Li 2010, p. 66). Mbembé (2003) emphasises, therefore, the importance of looking at the 'work of death' so its threat is better heeded as a prevailing technique of governance.

The stories of domestic violence and forced eviction told in *Home SOS* include both literal physical death and also social death, as well as trade-offs made between them by women and other actors. That domestic violence and forced eviction warrant placing in this life-threatening register is reflected in statements by the United Nations:

> In the most extreme cases, violence against women can lead to death. Two thirds of victims of intimate partner/family related homicides are women (United Nations 2015, np).

> In addition to being a violation of the prohibition on arbitrary or unlawful interference with the home, forced evictions all too often result in other severe human rights violations, particularly when they are accompanied by forced relocation or homelessness. For instance, if no adequate alternative housing is provided, victims of forced evictions are put in life- and health-threatening situations (United Nations 2014, p. 1).

Embracing a feminist geopolitical perspective which remedies how 'state and statecraft are treated as abstract forces that float above the contingencies of everyday lives and spaces' (Coleman 2009, p. 904), I bring the home into greater bio-necropolitical contention by considering women's killable homes and bodies. That the home is the most likely place globally for a woman to be killed (United Nations Office on Drugs and Crime (UNODC) 2018) renders this task especially prescient. While 'life and the home are so intertwined that it is almost impossible to think about one without the other' (Desmond 2016, p. 300), death also has an intimate embrace that requires belated exploration. Scholarship on domicide, 'the deliberate destruction of home against the will of the home dweller', is a rare exception in its direct focus on the 'murder of home' (Porteous and Smith 2001, p. 3). The destruction of home, Porteous and Smith argue, causes 'loss of historical connection; a weakening of roots; and partial erasure of the sources of memory, dreams, nostalgia and ideals' (p. 63). Uncomfortable with the finality of homes unmade in their book, and its lack of nuanced analysis on the impacts of home loss, I pursue a more processual, agentic, and gender-differentiated course (see also Nowicki 2014 for a critique).

In the interplay captured between life and death, *Home SOS* shows the variable disposability attached to Cambodian women and their homes by a government apparatus that places onus on marriages staying together in situations of domestic violence, and those that are 'just' collateral damage in the case of forced eviction. The former is deemed more biopolitically worthy, grievable, and less disposable than the latter, which are surplus to, and obstructive of, capitalist accumulation. Yet despite these differences between the management of domestic violence and forced eviction, both require women to take on the burden of 'upholding the work of death' (Mbembé 2003, p. 13) by (further) risking bodily integrity to ensure the security of others, be this a spouse, children, community, and/or the nation. Chapter 1 introduced, for example, how the enhancement and protection of Cambodian government interests and priorities are predicated on the sequestering of others, but particularly women's homes and bodies. Together, domestic violence and forced eviction demonstrate the (attempted) appropriation of bodies and homes that are variously constituted as surplus and intrinsic to accumulation. Only through the analytical collision of domestic violence and forced eviction is it possible to produce a nuanced account of women's homes, marriages, and bodies subjected to the differentiated *manoeu-vres* of, and grievable statuses given by, governmental actors.

Withstanding the 'fleshy, messy, and indeterminate stuff of everyday life' (Katz 2001, p. 711) being the constituent matter of the book, I provide pointed insights into the instrumentalisation of the home by the Cambodian government in order to push women to conform and submit to patriarchal and bio-necropolitical power. Government expression of sovereignty resides, in this remit, in the fantasy, capacity, and power to dictate who lives and who dies. Protracted situations and impacts of domestic violence and forced eviction can work to keep women in a permanent state of injury stuck in, or dislodged from, the home. The crisis ordinary becomes, in this context, a compensatory practice and status vacillating in between the home as a locus of life and death.

That domestic violence and forced eviction are shown in the book to be fundamentally connected to a women's relationship to the nation-state and their killable status functions as an important corrective to what Melissa Wright (2011, p. 726) critiques as 'the gaps in universalist depictions of the necropolitical and biopolitical forces at play in politics, economics, and culture'. From Wright's (2011) writing on femicide, to broader inter-disciplinary scholarship on honour killings (Ahmetbeyzade 2008), microfinance suicide (Roy 2012), women in war (Tyner and Henkin 2015), asylum claim-making (McKinnon 2016), and trans-sterilisation (Repo 2019), there exists a modest yet growing body of literature that problematises the neglect of gender (and sexuality) in discussions of necropolitics and the shortening of life (see also Alves 2014).

By engaging with precarity, a condition that 'is coming to be expected and/or accepted in a contemporary world defined by "crisis ordinary"' (Harris and Nowicki 2018, p. 389), I further explore how people are differentially exposed to injury, violence, and death. Precarity, Judith Butler (2009, p. 25) explains, is politically induced and designates that 'certain populations suffer from failing social and economic networks of support' and brings to the fore women's disproportionate acquaintance with harm. Although precarity has traditionally been considered in terms of formal workplace dynamics and neoliberal restructuring, my research data calls for a more expansive critical geography of precarity that more closely sutures matters of social reproduction, or *'life's* work' (Strauss and

Meehan 2015, p. 2, emphasis in original); that is: biological reproduction; unpaid production of goods and services in the home; social provisioning through voluntary work directed at meeting community needs; the reproduction of culture and ideology; and the provision of sexual, emotional, and affective services to maintain family and intimate relationships (Hoskyns and Rai 2007).

When it comes to mobilising the concept of precarity in this social reproductive context, Nancy Ettlinger's (2007, pp. 319 and 324) approach is sympathetic to my own in going beyond precarity as a bounded historical condition. She argues that 'beyond effects of specific global events and macro structures, precarity inhabits the microspaces of everyday life' and although no one escapes it 'one might argue that some people who experience more constraints than others also experience more dimensions of precarity'. Investigating the feminised responsibility for the home and its precarity as a form of gender-based violence, I push feminist studies forward by addressing a noted gap in literature on the household and social reproduction, and their political economy dynamics in relation to violence (Rai and Elias 2015). Cambodian women, I argue, are framed as naturally endowed with the capacity to manage all that is thrown (sometimes quite literally) at them, and that this is part of the ordinary crisis that women encounter in their survival-work. Expectations of personal resilience situated as necessary and as culturally appropriate in a Cambodian context mean, therefore, that women are disproportionally affected by domestic violence and forced eviction in both direct and indirect ways.

Home SOS demonstrates that while the home is implicated as an exclusionary and unhomely site of domestic violence and forced eviction in Cambodia, justice is simultaneously premised on (re)claiming the material and ontological security that it nominally affords. Yet it is this very vision of harmonious domestic life that is harnessed in governmental cultural logic and that exerts pressure on women especially to uphold home, and associated gender ideals, under challenging circumstances. In this cultural logic, the structural violence that women contend with is rendered visible. That the home is as much a symbolic as material entity is crucial in thinking through these geographies of violence. Symbolic violence, 'the power of suggestion which is exerted through things and persons' (Bourdieu 1991, p. 52) has the potential to be lived in, and through, its walls.

In the light of these nuances and inconsistencies, the book connects with long-standing feminist debates on the tensions of romanticising versus rejecting home as an ideal (Varley 2008). It exposes why the home is so important for feminists and at the same time so problematic to valorise. As Eleanor Jupp (2017, p. 350) summarises, home 'has, of course, been seen as the primary scene of women's oppression within much feminist analysis. Its realisation has been seen to depend on the subordination of women's own desires and projects, and exclusion from a public realm, as well as a place potentially of fear and violence.' But while domestic space has been traditionally attributed to oppressive gender norms that run the danger of enshrining women in a 'cult of domesticity' (Safa 1995, p. 42), my own scholarship questions unnuanced and uncritical perspectives that risk delegitimising women's sense of belonging, worth, and pride tied to their social reproduction of the family. A gender politics raised by the domestic sphere cuts both ways, with there being value in asserting the 'value of domestic cultures and women's creative shaping of them' whilst at the same time avoiding 'the risk of seeming to derive this tie from some kind of feminine "essence"' (Fraiman 2017, p. 16).

In this sense, the research in *Home SOS* indicates the dangers of foreclosing ideals for women to draw on in situations of domestic violence and/or forced eviction. This is because precarity 'is directly linked to gender and sexual norms since those of us who do not live our genders and sexualities in "intelligible" ways risk violence, discrimination, harassment and death' (Johnston 2018, p. 6). African-American feminists (Collins 1998; hooks 1991) have been especially influential in critiquing assumptions about the home as an oppressive place for women, given discrimination faced in the public sphere. Seminal scholarship by bell hooks (1991, p. 47) focuses on racialised processes of oppression through slavery and segregation and finds that it is 'in that "homeplace" most often created and kept by black women, that we had the opportunity to grow and development to nurture our spirits'. Ideals of home that express human values of preservation, safety, and privacy are thus to be defended (Young 2005). This is because for many groups, including 'the poor or transgendered person, the placeless immigrant or the woman on her own, aspiring to a safe, stable, affirming home doesn't reinforce hierarchical social relations but is pitched, precisely, against them' (Fraiman 2017, p. 20).

Taking these viewpoints on board, *Home SOS* goes on to show how domesticity is an important coordinate of public value that can represent and function as a burden, but also as a resource, for those in SOS situations. It has the potential to be simultaneously disempowering and liberating. Research in Mexico by Mario Bruzzone (2017) focuses on Las Patronas, a charitable organisation of women who throw home-cooked bags of rice and beans to migrants on passing freight trains driven north by poverty and violence. Their cooking not only fulfils gender roles and provides maternal authority, but also becomes a key ingredient in their push for social change. Gender ideals are thus strategically harnessed rather than simply conditioning of their experiences. Women's 'extensive domesticity' (p. 247) thereby becomes a spatial strategy and spatial analytic to forward outward looking and political goals. Explored together in the book, the empirical material on domestic violence and forced eviction reveals how women cope with, and innovate from below, under regimes of bio-necropower that render their homes and bodies materially, existentially, and socially precarious.

Intimate War and Slow Violence

In the previous section on bio-necropolitics and precarity, I emphasised the interplay between life and death that women encounter and the importance of these concepts for the empirical analysis that follows. In this next section I concentrate my focus on ideas of intimate war and slow violence in order to probe the attritional dynamics of domestic violence and forced eviction. In regard to the former concept, I argue that domestic violence and forced eviction are expressions of intimate war against women's homes and bodies located inside and outside of the traditional purview of formally declared war. Developing work that destabilises historic binaries between war versus peace (Gray 2018; Inwood and Tyner 2011; Pain 2015; Sjoberg 2013), the book reveals how the crisis ordinary represents an impasse between these binaries. If the dislocation of the everyday from geopolitics arises from peace being 'typically read as normal and war as aberration or disruption' (Cowen and Story 2016, p. 352), then my aim is to show that gender-based violence and the crisis ordinary complicates any such distinctions. Indeed, together, domestic violence and forced eviction unsettle and break down dualistic lines of thinking between war and not-war and forefront women's blurred encounters of them.

Research across geography and IR has a tendency, however, to ignore 'war' within the domestic context. As James Tyner (2012, p. 153, emphasis in original) notes of the former discipline, 'we fail to adequately question the *most common* forms of violence, the elephantine violence that occurs daily.' The gendered 'battlefield' of everyday life remains typically ignored, despite the fact that 'gender links violence at different points on a scale reaching from the personal to the international', and connecting 'battering and marital rape, confinement, "dowry" burnings, honor killings, and genital mutilation in peacetime; military rape, sequestration, prostitution, and sexualized torture in war' (Cockburn 2004, p. 43). 'Looking at *where women are* and *where gender is*' shows, therefore, 'that war, terrorism and insecurity are as often in the bedroom as on the battlefield, as often in the family home as in houses of government' (Sjoberg and Gentry 2015, p. 358 emphasis in original). Domestic violence and forced eviction reveal these spatial entanglements between war and the intimacies of everyday life.

Extending this line of thought, the study of intimate war is not just about reorienting the site of scholarly analysis to the home; it is being attentive to its enactment via, and to the detriment of, the home and its inhabitants. Given that localised instances of violence need to be thought of as part of a wider assemblage of space (Springer 2011a), the grounding of the book in the extra-domestic illuminates how domestic violence and forced eviction are marked by their rootedness within, and connections to, wider sites and processes of power. For example, Rachel Pain's (2015, p. 64) work contends that domestic violence and international warfare are part of a 'single complex' with intimate war denoting 'not a term for one or the other, but a description of both'. Intimate war, she argues, 'gains its devastating potential precisely because it does not concern strangers, but people in relationships that are often long term' (p. 67). Her research on the dynamics of domestic violence in suburban Scottish homes reveals similarities with modern warfare including tactics of shock and awe, hearts and minds, cultural and psychological occupation, just war, and collateral damage.

In *Home SOS*, I contend that domestic violence and forced eviction inflict these kinds of physical but also emotional wounds against women's bodies and social relationships in the home because of their long-term duration and legacies rooted in intimacies of knowing, loving, and caring. Intimacy relates strongly, though not exclusively, to the affective space of the home and the interiority of family life lived in it (Valentine and Hughes 2012). That crisis is 'an illogical departure' from a norm and 'serves the practice of unveiling supposed underlying contradictions' (Roitman 2013, pp. 11 and 9) brings to the fore the tension and chasm between the home as a site of intimacy and violence. My own adoption of the spatial metaphor 'intimate war' allows for the relationship between domestic violence, forced eviction, and war to be made more proximate through the analysis that follows. It is a 'countertopography' that 'demonstrates the rich insights to be gained by interrogating state-sanctioned violence via everyday, domestic, family, and community-based dynamics, as opposed to, or in connecting with, a traditional focus on formal, public, "P"olitical realms' (Faria 2017, p. 14; see also Little 2019).

Intimate war, akin to the crisis ordinary, instils recognition that violence and the survival-work it necessitates have become domesticated and routinised within women's 'ordinary' practices. This approach is an important one given that domestic violence and forced eviction are becoming lived as 'regularized warfare' (Mbembé 2003) in times of official peace – their intensification fuelled by the political and economic mechanisms of capitalism's dominance in cahoots with patriarchal control (Federici 2018b). Domestic violence, for example, is still tolerated by the courts and police as a legitimate response to women's noncompliance in their domestic duties – from child-rearing, food provisioning and preparation, laundering and cleaning – which should enable male freedom to participate in formal production outside of the household. Forced eviction, meanwhile, paves the way for 'new enclosures' by involuntarily removing women from their homes, land, and even marriages.

Mbembé's (2003, p. 27, emphasis in original) interpretation that sovereignty 'means the capacity to define who matters and who does not, who is *disposable* and who is not' thus warrants a spatial turn by questioning the 'where' of disposability in this intimate war. When places and bodies stand in the way of capitalist logics, violence and oppression are the likely outcome. Silvia Federici (2018b, p. 51) posits, for example, that escalations of violence against women rooted in the capitalist, patriarchal order, are essential to a 'new global war' against women 'because of what women represent in their capacity to keep their communities together'. This is especially pertinent to the home, normatively understood as a shared space of intimacy and belonging. Not only does Federici argue that the newly emergent political economy fosters more violent familial relations but also that a new surge of violence has roots in new forms of capital accumulation that involves land dispossession and the necessity to attack women in order to achieve this. As a result, 'precarity and resilience are the twin logics of a neoliberal order that abandons populations in pursuit of profit and then seeks to naturalize those abandonments as the only possible course of action' (Masco 2017, p. S73). *Home SOS* therefore roots and routes Federici's work on the global phenomena of war against women's bodies within, and through, the domestic sphere. These are embodied struggles that should be central, rather than peripheral, to debates on violence, patriarchy, and the deprivation of shelter. As Michele Lancione (2019a, p. 5) asks,

> Why are the efforts of millions of women fighting to live within their homes rele-gated to the rubric of 'empowerment' and 'capabilities', or registered only within the remit of feminist debates, rather than being seen as part of a quintessential fight to liberate housing from its patriarchal, masculine, violent ethos?

The joint study of domestic violence (law) and forced eviction (activism) reg-isters the fight for home within, but also beyond, feminist debates. For example, in *Home SOS* I develop scholarship on intimate war by arguing that attrition war-fare is a temporal register that also deserves exploration. Connecting literatures on intimate war with those on slow violence to explore this type of warfare com-plicates 'conventional assumptions about violence as a highly visible act that is newsworthy because it is event-focused, time bound, and body bound' (Nixon 2011, p. 3). Rob Nixon distinguishes slow violence as that which 'occurs gradu-ally, and out of sight, a violence of delayed destruction that is dispersed across time and space, an attritional violence that is typically not viewed as violence at all' (p. 2). If necropolitics 'is, in its most visible form, governing through death, slow violence is both its mode of operation and its effect at the level of the everyday' (Mayblin et al. 2019, p. 5). The crises of domestic violence and forced eviction are invested with the long-lasting harms of slow violence that are too often 'outside our flickering attention spans' (Nixon 2011, p. 6).

In a move to address such occlusions, scholars have become more attuned to 'sluggish temporalities of suffering' that 'dispossess without even breaching thresh-olds of eventfulness' (Ahmann 2018, p. 144). Work on crisis has been traditionally understood, in contrast, as a 'a privileged designation – a moment of rupture – that incites action and brings contradictions to light' (Ahmann 2018, p. 144). Domestic violence and forced eviction have 'eventful' moments, be this the punch of a fist or the crashing down of a bulldozer, but these are rarely singular moments as statistical data from Cambodia attests. They are more dispersed processes of dislocation that have tended to escape attention in comparison to civil and inter-national wars (Baviskar 2009). Slow violence compels, therefore, that 'we look beyond the immediate, the visceral, and the obvious in our explorations of social injustice' (Davies 2019, p. 2). For example, while there may be an initial disloca-tion and trauma, survivors sometimes must deal with long-term or gradual harms associated with protracted circumstances of displacement (Murrey 2016; Tanyag 2018) that can be conceptualised as slow violence (Hyndman 2019).

What the crisis ordinary, intimate war, and slow violence have in common is their insistence on the habitual and on-going harms of everyday life. As articulated by Ben Anderson et al. (2019, p. 2), concepts such as these respond 'to a fraying and breaking down of the geo-historical promise and hope that the everyday or ordinary can be separated from emergency/disaster'. They show how 'this promise and hope was only ever available to certain valued lives, came at a cost to other racialized, gendered, and classed lives, and has ongoing (after)lives embedded in the persistence of colonialism and other material and affective infrastructures of violence'. The twin study of domestic violence and forced eviction demands not only that these conceptualisations of everyday brutality are linked but that their gendered politics and outcomes are scrutinized in greater depth.

Law and Lawfare

Home SOS demonstrates how in the study of crisis ordinariness, bio-necropolitics and precarity, intimate war and slow violence, law is a significant 'splice' (Blomley 2003) in the story. Law seeps in and out of the home influencing power relations within and beyond it. It is part of the 'scenic tableau of bodily existence' in the crisis ordinary (Berlant 2014, p. 27). As Jeannie Suk (2009, p. 3) elaborates in *At Home in the Law*:

> In areas of utmost importance to individuals' relations to the state and to each other home is often overlooked as though it was self-evident and contained axioms from which legal results follow. But the legal meanings of home are ambivalent and contested. The home is a site of struggle over the most basic concepts that frame and construct our evolving legal universe.

Related to this point, a recent critical legal studies scholarship has made an effort to move away from technical legal analyses of property rights to focus on the human consequences of law related to the home (Fox O'Mahony and Sweeney 2016). While in legal geographic scholarship the home has been framed as a setting 'through which the ins and outs of a variety of power relations are established, enacted, revised, and reproduced' (Delaney 2010, p. 77), associated work has its limitations too, including a restrictive and disembodied onus on property rights. Emphasising how law is deeply implicated in multiple experiences of home, new socio-legal writing underscores the significance of law in both forging and mitigating precarity in the domestic domain (Carr, Edgeworth, and Hunter 2018). It suggests an important direction of travel in newer work that is attentive to lived experiences of the legally contingent nature of security afforded by home space.

Appraising the legal geographies project, as it is known, also brings about similar critiques and directions of travel. The project, emerging in the 1980s onwards, has lacked appreciation of the law's gendered manifestations and reverberations (Brickell and Cuomo 2019; Cuomo and Brickell 2019). *Home SOS* therefore contributes to forwarding feminist legal geography and understanding of how different types of laws, be these international, national, and/or customary, have the power to wage, as well as relieve, intimate war in women's lives. Socio-legal life, in this regard, 'is constituted by different legal spaces operating simultaneously on different scales and from different interpretive standpoints' (de Sousa Santos 1987, p. 288; see also Valverde 2015). The book thus explores the differential experiences of women living in pluri-legal fields of life and death and builds on law as a key medium by which feminist scholars have explored how subjects are produced. Linking to Michel Foucault's work, Judith Butler (1989, p. 601) contends that the 'body is a site where regimes of discourse and power inscribe themselves, a nodal point or nexus for relations of juridical and productive power'. Questioning the pregiven, in some senses there, body, she asks 'What shape does this body have, and how is it to be known?' How different bodies are produced in and through law is a key realm for exploring these provocations as part of bio-necropolitical power play on display in Cambodia. As Butler (1990, pp. 134–135) recognises, 'Law is not literally internalized, but incorporated, with the consequences that bodies are produced which signify that law on and through the body'. This is especially prescient given that law 'sits albeit sometimes unrecognizably, between the call for justice and the imperatives of violence ordering of the societies in which we live' (Sarat 2001, p. 13). For these reasons, juridical and cultural articulations of law and legality have a powerful influence in excluding, marginalising, and disciplining certain bodies (Brickell and Cuomo 2019).

Interest in the 'deadly embrace' of law and violence has brought attention to the associated concept of lawfare (Gregory 2007, p. 211; see also Doshi and Ranganathan 2019; Jones 2016; Jones and Smith 2015; Kittrie 2016; Springer 2013a). The Lawfare Blog[3] links the popularisation of the term with the US military figure Charles Dunlap (2001, np) who saw 'law as a weapon of war' and 'disturbing evidence that the rule of law is being hijacked into just another way of fighting (lawfare), to the detriment of humanitarian values as well as the law itself'. While 'rule of law' is generally understood as a value to be respected and a mechanism via which to guarantee justice and human rights to citizens, 'rule by law' – personified through the misuse conception of lawfare – is a distortion more easily conceived of as an instrument of oppression (Brickell 2016). Although it has been argued that 'rule of law and rule by law occupy a single continuum and do not present mutually exclusive options' (Holmes 2003, p. 49), the distinction remains nonetheless instructive. Furthermore, despite some additional kickback from military studies about the 'random' use and application of lawfare in certain disciplines, including geography (Voetelink 2017), it is a meaningful catalyst for bringing to the fore the political operation of law in the intimate wars that women are facing in their daily lives, but that continue to be side-lined. The domestic violence and forced eviction case studies therefore work, in tandem, to interrogate 'how law shapes, and takes shape through, multiple power-laden systems of domination that govern the lives of differently positioned women and bodies' (Gorman 2019, p. 1054).

Rights to Dwell

As Chapter 1 set out, domestic violence and forced eviction can be thought of as forms of displacement. Removal from home against the will of individuals, families, and communities is a gross violation of human rights, in particular the right to adequate housing (United Nations Commission on Human Rights 1993, p. 227). As the UN Special Rapporteur on Housing notes, 'The right to life cannot be separated from the right to a secure place to live, and the right to a secure place to live only has meaning in the context of a right to live in dignity and security, free of violence' (United Nations General Assembly 2016a, p. 11). Both domestic violence and forced eviction transgress and make imperative, therefore, the 'right to dwell' (Davidson 2009, p. 232). The right to dwell emphasises 'a right to inhabit the abstract space comprising "home" in a wider sense' than just its physical infrastructure (Hubbard and Lees 2018, p. 18; Baeten and Listerborn 2015). This spirit is echoed in the Committee on Economic, Social and Cultural Rights (1991), which states that:

> The right to housing should not be interpreted in a narrow or restrictive sense which equates it with, for example, the shelter provided by merely having a roof over one's head or views shelter exclusively as a commodity. Rather it should be seen as the right to live somewhere in security, peace and dignity.

Home SOS studies, accordingly, how domestic violence and forced eviction compromise one or both conceptual parameters of home as a physical location in which people reside and as an imaginative space of emotion and belonging (Al-Ali and Koser 2002; Blunt and Varley 2004; Blunt and Dowling 2006; Rapport and Dawson 1998). The book thereby concentrates on what happens when this occurs and how their reclamation is orchestrated and pushed back. It picks up, on this basis, from Louise Waite's (2009, p. 412) socio-political framing of precarity as 'both a condition and a point of mobilisation in response to that condition'. Therefore, while the crisis ordinary is felt and lived through the domestic sphere, its loss can also bring some hope in its creation of an 'impasse' (Berlant 2011, p. 5), a cavity in which possible capacity and outcome can grow 'to make a person or a people available for the next (potentially transformative, if maddening encounter)' (Berlant 2014, p. 32). The home in this guise represents a 'space to recalibrate, adjust, and coordinate feasible ways of moving forward' (Harris, Nowicki, and Brickell 2018, p. 156). In *Home SOS*, I critically assess domestic violence law (DV law) and forced eviction activism as two avenues for potentially building new 'infrastructures for reproducing life' free from violence in the home (Berlant 2011, p. 5). I also explore women's survival strategies as they sit variously alongside, outside, or in rejection of these interventions. The book therefore intervenes in debates connected to the scope and limits of citizenship, rights, and agency in the context of rights to dwell.

Traditional discourse on citizenship has shown a tendency to exclude the private realm from relevance (Yuval-Davis 1997) and neglect 'rights, obligations, recognitions and respect around those most intimate spheres of life' (Plummer 2001, p. 238). To counter this, the book builds on feminist geography work that has emerged in the last five years on marriage as central to notions and articulations of family and citizenship (Ansell et al. 2018; Brown 2015; D'Aoust 2018; Waitt 2015; Wemyss, Yuval-Davis, and Cassidy 2018; Yeoh, Chee, and Vu 2014). As a result, I provide a 'more complete geography of citizenship' by exploring the 'hidden spaces' of home life and marriage (Staeheli et al. 2012, p. 641), which are too often discounted.

The spatial pivot of the book on the extra-domestic also underscores the centrality of home as a 'gateway right' (Amnesty International 2012). The domestic violence and forced eviction material presented in *Home SOS* shows that without the right to dwell, other rights, such as living safely, are much harder to achieve. This matters, given that the home is not only a place in which people live but also 'an idea and imaginary that is imbued with feelings' (Blunt and Dowling 2006, p. 2). As David Madden and Peter Marcuse (2016, p. 12) write:

> It [the home] is a universal necessity of life, in some ways an extension of the human body. Without it, participation in most of social, political, and economic life is impossible. Housing is more than shelter; it can provide personal safety and ontological security. While the domestic environment can be the site of oppression and justice, it also has the potential to serve as a confirmation of one's agency, cultural identity, individuality, and creative powers.

The loss of home, be this materially and/or metaphorically, can therefore manifest as a form of 'compound disempowerment' (Millar 2015) that women and their families experience and try to counter through survival-work. Anne Allison (2013) points out in her research on the insecurity of work in Japan, that precarity which begins in one arena can easily slip to others. Precarity cross-cuts women's lives, spirals even, and with this stretches beyond the location and duration of any original violence. It has been argued, for example, that the impacts of domestic violence on women impinges on their 'housing, mental health, and employment as a key aspect of citizenship' (Zufferey et al. 2016, p. 1). The United Nations (2014, p. 1) has recognised too that in forced eviction cases, severe trauma can result and 'set back further the lives of those that are often already marginalized or vulnerable in society' (notably women). Domestic violence and forced eviction therefore exacerbate the disadvantaged position that women often have in relation to men when it comes to citizenship, despite claims to universalism (see Lister 2003). Violences of attrition they face therefore extend to the chipping away of citizenship rights in both instances.

Home SOS thereby develops geographic scholarship that critically interrogates the concept and utilities of rights and 'the particularities of place, the spatialities of power and difference, and the scalar interplay between universal ideals, the territoriality of the state, and the experiences of individual bodies' (Laliberté 2015, p. 58). Its domestic tilt links with, and sheds lights on, a human rights regime that has grown internationally to encompass a variety of private threats to human dignity. The book provides insights into the historically, geographically, socially, and culturally contingent nature of rights and citizenship in Cambodia today. It therefore offers a window on to vernacularisation: 'the extraction of ideas and practices from the universal sphere of international organizations, and their translation into terms that make sense to their local communities' (Merry and Levitt 2017, p. 213). Set in a country where traditional norms dictate that domestic problems are not communicated beyond the spatial confines of the home, I demonstrate, in this vernacular frame, how personal and collective action has tried to reframe domestic issues as societal problems through rights-based discourse. However, at the same time I reveal how to be Khmer, as it is exerted by government spokespersons, is to reject the modern pantheon of human rights (see Chapter 3). 'Active citizenship' in the sense of bringing 'private' rights violations into public view is considered by ruling party organs not to be in the national interest. This is despite the country being signatory to a swathe of international human rights treaties. Government resistance to 'reform that would put liberal legal frameworks and human rights ideologies into practice' (Morris 2017, p. 29) is a perennial issue in Cambodia. Collective order is prioritised over personal freedom, and respect for strong leadership is combined with a sustained attachment to family and conventional patterns of authority (Stivens 2006). The harmony and stability of Cambodian marriages and households – women's affective labour in the practice and performance of the crisis ordinary – is therefore considered mandatory in the reification of the secure nation.

While in the case of domestic violence this translates into keeping 'fire' in the house and eschewing legally sanctioned rights, in forced eviction cases this would mean letting the destruction of homes go unchallenged. Under such circumstances, it has been argued that it is important to distinguish between active citizens and activist citizens, the former depicting the citizen as performing given scripts and the latter writing scripts that have an effect (Isin 2009). I would caution, however, that this bifurcated vision of citizenship as performing either given or invented scripts misrepresents the multiple and contradictory scripts which are operative in Cambodia. As I have previously written of gender scripts more specifically (Brickell 2011, p. 458), 'although the image of the present-day woman still carries with it many ideas about acceptable femininity, the multifaceted influences now operating in Cambodian society have created a situation in which there is no singular ascription process at work' (see also Lilja 2017). The combined study of domestic violence and forced eviction allows for the selective instrumentalisation of different visions and intensities of rights in time and space to be brought into greater view.

In this vein, *Home SOS* challenges the conventional contours of what constitutes activism and brings women's helping-seeking in domestic violence cases into the orbit of what counts as activism. Concerning forced eviction, I concentrate on an 'activist' group of self-labelled housewives who contested their displacement for over a decade through public visibility and political agitation (see Askins 2011; Horton and Kraftl 2009; Pain 2014; Chatteron and Pickerill 2010 on everyday activism). Women's diverse forms of activism, and the violence they can consequently encounter, brings into further contention the point made by Joe Turner (2016, p. 143) that efforts to (re)politicise the analysis of citizenship 'by bringing into focus those acts which "protest" exclusory regimes' has simultaneously helped to 'delineate the contingency of citizenship itself'. Both examples, of DV law and forced eviction activism, underscore the gendered contingency of citizenship and the direction of the material to come in the book on the survival-work that women undertake.

Conclusion

In this conceptually led chapter, I have explored ideas of crisis ordinary and survival-work, bio-necropolitics and precarity, intimate war and slow violence, law and lawfare, and rights to dwell. It has encouraged scholastic conversations across these multiple theoretical arenas and has drawn attention to the need for crisis and the capacity to dwell to be interrogated through different theories of violence, precarity, and demise. With the exceptions of precarity and intimate war, these conceptual ideas have not originated from feminist work, but have increasingly been the subject of feminist substantiation as well as critique. *Home SOS* continues, and extends, this work through its focus on domestic violence and forced eviction as embodied entanglements with patriarchal and capitalist forces. Both are forms of gender-based violence that are articulated in, but also promulgated from outside, the extra-domestic home.

They are also everyday encounters, which for many women in Cambodia are a chronic part of ordinary life. These encounters amass through increments and take an especially acute toll on women and the intensity of survival-work necessitated. The temporal chronicity of crisis inherent in domestic violence and forced eviction is why conceptual notions such as the crisis ordinary and slow violence hold such potential purchase in the study of gender-based violence. Through these complementary conceptual ideas and the integrated study of domestic violence and forced eviction, the rest of the book aims to examine these intimate geographies of violence and the ever-dynamic relationship between gender, space, and power in twenty-first century Cambodia. The contextual scene is set up next in Chapter 3, which provides the backdrop in which these domestic crises are produced and reproduced in time and space.

Endnotes

1 The World Health Organization's (2002, p. 4) definition of direct violence is: 'The intentional use of physical force or power, threatened or actual, against oneself, another person, or against a group or community, that either results in or has a high likelihood of resulting in injury, death, psychological harm, maldevelopment or deprivation.'
2 Please note, Mbembé's 2019 book was not published when my book manuscript was submitted.
3 www.lawfareblog.com.

Chapter Three
National Trajectories of Crisis in Cambodia

Introduction

In this chapter I provide the contextual background to *Home SOS*. I then chronologically outline some of the key socio-economic and political dynamics that are important for understanding contemporary Cambodia and its crises over time. I am particularly attentive to matters of marriage and family life, violence against women, and forced eviction, given the focus of the book. Because the history of the country has been extensively written about, my aim is to provide a short introduction that offers the necessary contextual knowledge. This basis is important in a country that has seen the 'extreme compression' of many sources of change into a short period of time (Hall, Hirsch, and Li 2011, p. 2). To this end, I finish the chapter by exploring two tropes about the country that are enfolded into later arguments: the first, its supposed battlefield to market-place transition and, the second, political discourses of order and stability. Taken together, these sections provide the backdrop in which the crisis ordinary takes shape.

National Trajectories

This chronologically organised section of the chapter is divided into four: the first on pre-revolutionary Cambodia (pre-1970); the second on its closing to the world with the Khmer Rouge (1975–1979); the third on The People's Revolutionary

Home SOS: Gender, Violence, and Survival in Crisis Ordinary Cambodia, First Edition. Katherine Brickell.
© 2020 Royal Geographical Society (with the Institute of British Geographers). Published 2020 by John Wiley & Sons Ltd.

Party of Kampuchea (1979–1989); and the fourth on the country's reopening to the world in the decades since.

Pre-revolutionary Cambodia (Pre-1970)

For Cambodians of all ages, '"traditional" refers to pre-1970 customs; the way things were done "before the war"' (Luco 2002, p. 7). Judy Ledgerwood et al. (1994, p. 516) warn, however, that 'while it is heuristically useful or necessary even to use pre-Revolutionary Cambodia as a "baseline" to discuss post-1970s change in Khmer existence both in homeland and elsewhere, it is important to understand "Cambodian culture" as an intellectual construct that has long undergone transformations wrought both through endogenous and exogenous forces'. Once a mighty empire, spanning Siam (now Thailand), Burma (now Myanmar), Laos, and parts of Vietnam in the twelfth century of the Christian era, Cambodia shrank to a vassal state of both Siam and Vietnam until it was 'rescued' by France in the mid-nineteenth century and emerged from colonial rule as a modern, independent nation state in the 1950s (Brown 2000, p. 11). French rule viewed Cambodia as a backward, unchanging country, her people docile, immune to modernisation, and in need of the mission civilisatrice that France was also undertaking in Vietnam and Laos (p. 22). As a result, the French brought in Vietnamese bureaucrats to staff the civil service, causing anti-French feeling to grow among Cambodians, who objected to their presence. One of France's enduring legacies to Cambodia was also the 1884 Land Act, which legitimised private ownership of land for the first time and created a deterioration in the conditions for smaller farmers (Ovesen et al. 1996).

By the Second World War, French control in Indo-China was weak and pro-independence movements in the region were growing in strength, resulting in Cambodian independence in 1953. Independence did not quell Cambodians' fears of territorial and political domination, and political leaders were quick to exploit such fears, creating deep-seated prejudices against 'outsiders' such as the Vietnamese (Brown 2000). Before 1970, the Cambodian underground communist movement led by Pol Pot was fragile and did not pose any threat to the government of Prince Norodom Sihanouk. Sihanouk had dominated Cambodian politics since the French crowned him King in 1941, manoeuvring for independence from France and in 1955 abdicating the crown to become chief of state (Ledgerwood 2003). Whilst after independence there was no return to the glorious past of the Khmer Empire with Angkor as its material legacy (Winter 2004), Cambodians rejoiced in the presence of Sihanouk whose political manoeuvrings helped to bring foreign rule to an end (Ledgerwood 2003). Sihanouk's good relations with North Vietnam, compounded by his suppression of internal political dissent, held the Cambodian communist forces in check. Moreover, the communist movement could not win over the support of the masses because of the prince's popularity.

By the late 1960s, however, Cambodia faced serious economic difficulties provoked by a combination of factors, including Sihanouk's rejection of American aid, corruption, and the fact that by 1966 more than a quarter of Cambodia's rice crop was being sold illegally across the border to war-torn Vietnam (Ledgerwood 2003). This cut deeply into government revenues, which were dependent on export taxes on rice (Chandler 1991). The national education programme that Sihanouk established after independence in 1954 meant that there was a growing educated populace with expectations of social mobility. Antagonism towards the Sihanouk regime increased in the 1960s among the urban elite, students, and intellectuals, and the prince was overthrown in 1970 by General Lon Nol and senior military officials in a *coup d'état*.[1] While the United States denies orchestrating the coup, General Lon Nol's administration was immediately awash with US funding. The knowledge that US support would be forthcoming was undoubtedly an important motivating factor for the men who directed the coup (Ledgerwood 2003).

In 1970 US and South Vietnamese armed forces launched heavy air and ground military campaigns against North Vietnamese soldiers inside Cambodia. Their goal was to capture the headquarters of the Vietnamese communist movement, which was based inside Cambodian territory, but which was never found by the invading forces. The military offensive pushed the North Vietnamese soldiers deeper into Cambodian territory. By the end of 1973, the total bombs dropped on Cambodia reached 539 129 tons – three times more explosives than were dropped on Japan during the Second World War (Ablin and Hood 1988). Sihanouk, while in exile and with the encouragement and support of China and North Vietnam, formed a united front with the Cambodian communists to fight against the US-backed government in Phnom Penh. With growing support from the Cambodian people, arms from China and North Vietnam, anger over US bombardment, and appeals from Sihanouk to join their cause, opportunities were created for the Khmer Rouge. Led by the enigmatic Pol Pot, their armed forces were built up from around 800 soldiers in 1970 to a well-organised and well-disciplined force of 40 000 soldiers in 1973 (Ablin and Hood 1988). By this time the Khmer Rouge controlled most of the Cambodian countryside and over the next two years they advanced and finally took control of Phnom Penh on 17 April 1975.

The Khmer Rouge Regime (1975–1979)

At Thirteen, the nascent adult in me realizes that Cambodia is a nation that houses the living dead. Around me there are starving, overworked, and malnourished people. Death is rampant, as if an epidemic has descended on the villages… Death is like leaves in the Autumn, readily falling from the soft touch of the wind. I wonder who in my family will be the next victim (Him 2000, pp. 240–241).

The Khmer Rouge regime (1975–1979) is widely recognised as one of the past century's worst crimes against humanity (Becker 1998) and there exists considerable writing on its extraordinary violence in survivor accounts, including by Chanrithy Him quoted here (see Him 2000; Lafreniere 2000; Ngor 1987; Ung 2000; and Wagner 2002 as typical examples), and in academia (Chandler 1991,1992, 1996,1999; Cook 2006; Hinton 2005; Jackson 1989; Kiernan 2002, 2003, 2008; Mysliwiec 1988; Tyner 2008; and Vickery 1984 to name but a few). The policies of the Khmer Rouge were aimed at radically transforming Cambodia into a new society breaking completely with its past and for its populace to become 'pure' Kampuchean revolutionary worker peasants. The 'upper organisation' or *Angkar Loeu* was the governing body of the renamed Democratic Kampuchea (DK). The identity of its leaders, however, were hidden from the public. This secrecy was viewed to be essential, because it had helped them in the past and because enemies were allegedly attempting to sabotage the revolution (Ledgerwood 2003). Personal freedom was totally controlled by Angkar, 'the pineapple', so-called because it had eyes everywhere checking these perceived threats (Eng and Hughes 2016).

Immediately after the fall of Phnom Penh in April 1975, Pol Pot and his cadres began evacuating approximately three million people from towns and cities throughout the country to the extent that while other revolutionaries had 'distrusted cities ... only the Cambodians have emptied them altogether' (Stretton 1978, p. 118). In revolutionary eyes Phnom Penh had become 'an extreme case of an exploitative, debauched, worse-than-useless parasite of a city' brimming with lazy, unproductive, and politically misguided refugees (Stretton 1978, p. 118). Although Khmer Rouge cadres claimed that the evacuations were to prevent epidemics and starvation and to protect civilians from American bombing, this decision was, in fact, 'a calculated, political decision, part of a wider agenda with economic and ideological rationale' (Chandler 1991, p. 247). The evacuation's purpose was therefore to ensure Democratic Kampuchea's control over the urban population and to turn what they deemed the apparently culturally corrupt, economically, and politically exploitative urban class into a new and productive people. Using forced migration as a strategic means to disorientate and dislocate, the transportation of these populations from one end of the country to another was to become a 'hallmark' of the revolution (Becker 1998, p. 232). The memoir of Var Hong Ashe (1988, pp. 30 and 37) speaks to urban dwellers' experiences of being forced to leave their homes without any certainty of return:

A general air of misery hung over the whole crowd as we trudged along. Our entire fabric of life had been torn apart and, as I stared at the sea of unknown faces, my own uncertainty was mirrored in the looks which I received back. I had never seen the streets so packed with people before. Some, like us, were pushing cars; others pushed carts or bicycles with packages piled high behind the seat, so heavy that it seemed as if only the denseness of the crowd prevented them from falling over.

Others, who were poorer, and those who had been evacuated with little or no warning, carried bags on their heads or on the ends of bamboo poles slung over their shoulders. Some walked only with a bunch of keys in their hands, hoping to go back … In the history of the whole world, we had never heard of a revolution which had emptied the capital and which had not allowed the population to return. Why should they keep us away from our homes?

From a mother, her children, and possessions having been hastily loaded on to a cyclo as the Khmer Rouge took hold on 17 April 1975 (Figure 3.1), the forced displacement of the regime speaks to the imaginative as well as physical significance of home as a central rationale behind its defilement. Moving the urban population into the countryside in Cambodia enabled the Khmer Rouge to control every segment of the population and reduce people's ability to resist their new leaders (Chan 2003). It is an event which ranks, as 'one of the largest population movements in history' (Eng and Hughes 2016, p. 160) and the nulling and voiding of any rights to dwell. The former inhabitants of cities and towns were forced to engage in agricultural labour in the countryside. Those who could not transform, chose not to, or who were considered a threat to the revolution were imprisoned or eliminated. Khmer Rouge cadres, for example, immediately executed former high-ranking government officials, businessmen, and military officers.

Figure 3.1   The fall of Phnom Penh, 1975. Source: © Roland Neveu (RNBK.INFO). Reproduced with permission.

Traditional notions of social hierarchy were thereby reversed (Ledgerwood 1990) and personal narratives written by urban Khmer who survived the DK regime are testament to this (Him 2000; Ngor 1987; Ung 2000). Members of the wealthy urban elite who had possessed every advantage found themselves treated as the lowest rank of society. Like their Maoist counterparts in China, the Khmer Rouge leaders emphasised manual labour over knowledge. They claimed 'rice fields were books, and hoes were pencils', rendering teachers and those educated to the lowest rank of society. In the view of the Khmer Rouge, Cambodia did not need an education system. The Khmer Rouge leaders deliberately destroyed the foundations of a modern education. People with higher education, such as doctors, lawyers, teachers, professors, and former college students, were killed or forced to work in labour camps. The Khmer Rouge also engaged in the physical destruction of institutional infrastructure for higher education, such as books, buildings, and other educational resources. It is estimated that by the end of the Khmer Rouge time, between 75 to 80% of Cambodian educators had been killed, died of overwork, or had left the country (Ledgerwood 2003) and that, overall, 1.5 million people were killed. Life was extremely precarious, so much so that 'even the life that was spared was imprinted with death' (Um 2015, p. 46). This subjugation of life to the threat of death thus aligns closely with Mbembé's (2003, p. 14) conception of necropower as 'the generalized instrumentalisation of human existence and the material destruction of human bodies and populations' (see also Tyner and Rice 2016).

Pre-revolutionary institutions were also uprooted and old traditions, thoughts, and ways of life were forbidden in this intimate war. This included the eradication of the entire legal system.[2] Difference was deemed to be evil and Buddhism and its vestiges, being at the core of Khmer ideas of social hierarchy, were abolished. Homes, and even entire villages, were also physically demolished to make way for communal buildings and working teams replaced families as the new basic socioeconomic unit of society (Ebihara 1993). The Khmer Rouge viewed the home and family, 'as the most potent, hence the most feared, of all relationships of the former society' (Becker 1998, p. 211) and sought to erase them. The family was therefore reconfigured as a collective entity, with political figures assuming the role of parents and families physically separated by forced evacuation and extremist social policies. While historian Michael Vickery (1984) argues that the Khmer Rouge employed no deliberate policies to destroy the family, majority thinking is that its policies were designed for this very purpose (Ebihara 1990, 1993; Ledgerwood 1990; Ponchaud 1989).

During this time, women lost control of their primary areas of recognised authority concerning the household and its finances. Having said this, although the conflict imposed economic hardships on women, it also opened new opportunities to participate in the economic sphere because of the mass mobilisation of men into the military, mass killing, and increased labour demand for war and rehabilitation work (Kumar, Baldwin, and Benjamin

2001). This led to the undermining of the traditional sexual division of labour that had characterised Cambodian society (ibid 2001). Traditionally, for example, male work in Cambodia included ploughing, felling the jungle, hunting, woodworking, and house building – as well as statecraft and formal religion – while the female domain included transplanting, harvesting, vegetable growing, food preparation, weaving, pottery-making (in most areas), and marketing, as well as communing with ancestors and mediating with the spirits (Reid 1988).

During the Khmer Rouge regime women not only worked in the fields (Figure 3.2), but as militia who were mobilised in defence of rural areas on both

Figure 3.2   Khmer Rouge women's unit harvesting rice. Source: Courtesy of Documentation Center of Cambodia.

sides of the civil war – the first opportunity in recent Cambodian history that women had to exercise power in the public realm (Jacobsen 2003). The Khmer Rouge were also the first to organise women at the grassroots level, establishing a women's wing of the Communist Party of Kampuchea in the late 1960s (Kumar et al. 2001). This uniformity between men and women is expressed in one of six revolutionary songs, 'The Beauty of Kampuchea':

> O Beautiful, beloved Kampuchea, our destiny has joined us together, uniting our forces so as not to disagree. Even young girls get up and join the struggle (Translation, cited in Kiernan and Boua 1982, p. 237).

John Marston (2002) has pointed out that unlike pre-revolutionary Cambodian songs, these post-1975 versions did not distinguish between sections to be sung by men and others to be sung by women. This direct involvement of women in the revolution connects with the realisation in conflict studies that women's involvement is not necessarily a passive one (El-Bushra 2000). In Democratic Kampuchea, for example, women's capability to participate in violent conduct like men, or broadly perform the same tasks, is reflected in the constitution. Article 13 states:

> There must be complete equality among all Kampuchean people in an equal, just, democratic, harmonious, and happy society within the great national solidarity for defending and building the country together. Men and women are fully equal in every respect.[3]

The new participation of women was accompanied by their seeming enforced 'defeminisation' characterised, inter alia, by uniform haircuts in the form of a neat bob just below the ears, loose clothing, and the banning of jewellery. The reasoning behind this lay in the equation made between long hair and fashions such as flared trousers and mini-skirts and the lax morality and corruption of prior periods (Jacobsen 2003). Although there were female leaders among the revolutionary government, and the official party line endorsed gender equality, power remained in the hands of those connected to men in key positions.

Women were also the victims of multiple forms of violence, and arguably suffered the worst effects in terms of emotional and physical health, including miscarriage and enforced separation from children. Indeed, the country has been described as having a 'buried history of sexual crimes' including rape, sexual mutilation, and forced marriage (Anderson 2005, p. 789). An undetermined number of women suffered sexual violence during the Khmer Rouge regime, yet the prevailing myth was until very recently that rape did not occur during this time. This falsehood is today being shattered by a mounting body of testimonial evidence from multiple Women's Hearings on Sexual Violence Under the Khmer Rouge Regime, which provided the platform for survivors to speak out in public. An excerpt from the account of rape survivor Net Savoen is illustrative in this respect:

> When they arrived at the execution site in the forest, they started to rape the women and beat them to death with axes and finally they cut their throats. The pretty women were raped by them as they wished. Some women were raped by three to four men before they were killed. She could clearly see everything that happened because it was full moon. They started to kill and rape the women from the time the moon was rising at sunset until the moon was fully up in the middle of the sky. Then it was her turn. She was the last person among all who was standing and waiting for her turn until her whole body was numb because she did not know what to do (cited in Savorn 2011, pp. 35–36).

Theresa De Langis et al. (2014, p. 14) argue that forced marriages during the Khmer Rouge period should also be framed as forms of 'sexualized gender-based violence' given that 'rape was normalized and perpetrated with impunity, especially within marriage and for punishment'. On the wedding night, for example, Khmer Rouge spies could be found covertly hiding to check marriages were consummated. The legacy of forced marriages remains to the present day with the exploratory study by De Langis et al. finding that although some couples have remained in a relationship, half of them are dysfunctional and women report continuing spousal abuse (see also Deibler 2017).[4] As such, intimate war has not ceased, rather it has been carried into the intimacies of post-conflict life.

The People's Revolutionary Party of Kampuchea (1979–1989)

In early 1979 Cambodian attacks on Vietnamese border towns and villages, and reports of the decimation of the Vietnamese community, led to a force of 120 000 Vietnamese entering Cambodia, reaching Phnom Penh and ousting the Khmer Rouge (Brown 2000). Democratic Kampuchea (DK) was replaced by the People's Republic of Kampuchea (PRK) who ruled throughout the 1980s. With the collapse of the DK regime, hundreds of thousands traversed the country in search of family members and on their own came together and recreated cities and towns (Gottesman 2003), initially elated as they realised they could 'go home' (Daran Kravanh, cited in Lafreniere 2000, p. 154). This mobility of the population and the social chaos that ensued after the collapse of tight DK control resulted in low rice production in 1979. Thousands of people, especially elderly people and children, died of starvation.

Cambodia's judiciary was restarted after the Vietnamese takeover but was slow in its rebuilding of the shattered legal system and profession. In 1979 there are believed to have been only ten law graduates left in the entire country and basic legal rights, such as the right to defence, remained non-existent throughout much of the 1980s (Duffy 1994). At the same time, outside powers once again took sides and Cambodia 'reverted to a second, more low-intensity albeit still massively debilitating civil war between the PRK government and the alliance of anti-PRK guerrilla groups operating out of refugee camps on the Thai border' (World Bank 2006, p. 5). This conflict continued throughout the 1980s with Cambodia remaining closed to most of the world, except for the presence of some aid agencies (Neupert and Prum 2005).

Re-opening Windows to the World (1989 Onwards)

In 1989 Vietnam withdrew its troops. Thereafter, on 23 October 1991 the four warring factions – FUNCINPEC, the Khmer Rouge, the Khmer People's National Liberation Front, and the People's Republic of Kampuchea/State of

Cambodia – agreed to end their armed conflict and signed a peace accord known as the Paris Peace Agreements.[5] The agreements invested Cambodian sovereignty in the Supreme National Council (SNC), which was headed by Norodom Sihanouk and contained representatives from the four factions. The SNC in turn delegated all powers necessary to implement the accords of the United Nations Transitional Authority in Cambodia (UNTAC), whose mandate was far-reaching. To coordinate the repatriation of Cambodia refugees and displaced persons and to organise and conduct free and fair elections, UNTAC deployed some 16 000 military personnel and 5000 civilians, including 3500 police officers. Cambodian NGOs have also grown exponentially since 1991, enmeshed nonetheless in long-standing patterns of patron–client relations through their sponsorship by local patrons, dependency on foreign funding, and need to meet international donor imperatives (Coventry 2017).

The Paris Peace Agreements were not completely fulfilled with the disarmament process failing. The Khmer Rouge pulled out of the electoral process and continued to wage war after the election in May 1993. Although FUNCINPEC won the elections and Prince Norodom Ranariddh of FUNCINPEC (one of Sihanouk's sons) became First Prime Minister, Hun Sen of the Cambodian People's Party (CPP) assumed the position of Second Prime Minister. The coalition ran into trouble after 1996 and met an abrupt end when Hun Sen staged a coup against his co-premier in early July 1997. After some forceful international intervention, however, the country managed to hold an election on 26 July 1998. This time the CPP won, though arguably through political intimidation and massive fraud (Ledgerwood 2003). For some analysts therefore, 'we are not looking at a transition toward liberal democracy that has been stymied, derailed, or thrown into reverse, but rather one that has never even left the station' (McCargo 2005, p. 107).

Now one of the world's longest-serving prime ministers, Hun Sen has used his political power since 1985 to steer the country towards a neoliberal path. This has comprised rapid deregulation, privatisation, and liberalisation to attract foreign investment and aid (Springer 2009, 2010). These political economy turns have also had important implications for rule of law. As Alexandra Kent (2016, p. 15) observes:

> Establishing the rule of law is assumed to mark a break with the past and generate a culture in which state actors may not transgress the law for political gain. Using the Cambodian case … demonstrate[s] that instead of fostering self-sustaining and inclusive peace, the imperative for governments to liberalise their economies, and maintain stability may encourage formerly militarized elites to monopolise capital and use it to intimidate populations, perpetrate injustice and create social exclusion.

Cambodian leaders have therefore 'confounded the efforts of the international community to promote rule of law' given that this requires 'more than training legislators, judges, and the police forces; it is a transformative process that changes how power is distributed and exercised in society and as such it is inherently threatening to the power holding elite' (McCarthy and Un 2017, pp. 100 and 103). As Catherine Morris (2017) notes, liberal conceptualisations of the rule of law and its limitations on power do not hold sway in Cambodia where lawmaking is characterised by the inversion of justice, instilling of structural violence, and deepening of inequality. Kheang Un's (2019, np) recent analysis of Cambodia's 'return to authoritarianism' has also underscored how 'any judicial reform tends to promote "rule by law" or "weaponizing laws" rather than rule of law' in order to 'supress political opposition and civil society groups'.

The holding of power and Hun Sen's political leadership at its centre, encountered little challenge given that the impact of opposition parties had been limited to urban areas in successive elections (Hughes and Eng 2019). Perhaps for this reason, in the 2000s and early 2010s print media in Cambodia was relatively unconstrained and foreign human rights organisations were free to operate. The early 2000s also saw grassroots resistance build with the assistance of these international and local NGOs (Young 2016). The 2013 election brought with it the serious possibility of regime change for the country and a sense of hope was pervasive (Brickell and Springer 2017). As Astrid Norén-Nilsson (2019, p. 84) notes, 'cracks in the established political system' were evident at this time. In the end, those in favour of the opposition Cambodian National Rescue Party (CNRP) fell just short of securing an electoral victory, but the narrow margin of the ruling CCP's win concerned Hun Sen 2017. 'Like the monsoons, repression in Cambodia comes on a regular schedule, and the current season looks like it will be an especially long and stormy one', Sebastian Strangio (2016, np) wrote in its election aftermath.

Since the 2013 elections, Cambodia has seen a 'hardening authoritarianism' (Norén-Nilsson and Bourdier 2019, p. 3). Repression has taken multiple forms, including the defensive reaction to mass strikes and demonstrations involving hundreds of thousands of workers and which led to multiple deaths (Arnold and Chang 2017; Lawreniuk 2019; Lawreniuk and Parsons 2018). The crisis ordinary of repression was also particularly intense in the lead-up to the 2018 election. First, there has been a growing use of law as a tool to target social movements and the operations of NGOs (Curley 2018). The Law on Associations and Non-Governmental Organisations (LANGO) makes both national and international NGO registrations compulsory and only possible through governmental approval (see Royal Government of Cambodia (RGC) 2015). It was ratified in 2015 despite local and international consternation that it violates Cambodia's international law obligations and constitution (Morris 2017). Its coming into being is not necessarily surprising since NGOs are often perceived as agents of foreign interests and even oppositional forces against the government (Coventry 2017). Second, anti-government critics have been incarcerated or more permanently silenced. Kem Ley, a political commentator, for example, was gunned down in broad daylight on 10 July 2016 in what many have characterised as an assassination (see Norén-Nilsson 2019 on Kem Ley's fearlessness and legacy). The main opposition party, the Cambodia National Rescue Party (CNRP) was also dissolved in November 2017 and its leader, Kem Sokha, imprisoned soon after (Allad and Thul 2018, np). Journalist Alex Willemyns (2016, np) writes of these threats against dissenters:

> Arrest an activist. Slap them with a prison sentence far longer than many given to convicted murderers, rapists or drug traffickers. Then let them languish behind bars while pursuing concessions from the opposition. The open threat of corrupt courts being used against government critics should silence the most skittish among them. If there are any calls to release the prisoners, flatly deny control of the courts and cite 'the rule of law.' The use of such double messages – scaring political opponents with jail time while maintaining a position of strict constitutionalism when it comes to their release – has worked wonders for the CPP over the years, often leaving the opposition lame.

Third, alongside impairments to freedom of expression, association, and assembly, violent threats have been made to dampen the potential for dissent. A Human Rights Watch (2017a, np) survey of 2017 events included the following sketch:

> In May 2017, Hun Sen stated he would be 'willing to eliminate 100 to 200 people' to protect 'national security,' and suggested opposition members 'prepare their coffins.' On August 2, Minister of Social Affairs Vong Sauth said that protesters who dispute the 2018 election results will be 'hit with the bottom end of bamboo poles' – a reference to a torture technique used during the Khmer Rouge regime.

This political rhetoric reaffirms the importance not only of thinking through the bio-necropolitical parameters of Cambodian society of the past but also the recasting of intimate war in the violent bodily provocations of today. Such behaviours are representative of the sharp downturn in civil and political rights in 2017, which also saw the forced closure of several independent media outlets (Human Rights Watch 2017a).

On 29 July 2018, Cambodians headed to the polls for an election that has been criticised by the United Nations and Western countries as fundamentally flawed, given the enforced dissolution of the main opposition party. The United States and the European Union responded to the crackdown by withdrawing financial support and monitors from the election, a step followed by independent local and international NGOs that had overseen previous elections. With electoral credibility undermined, Hun Sen threatened citizens who did not vote. Voters therefore faced a difficult choice between casting a vote for one of the twenty minor parties or boycott the ballot and risk consequences from local ruling party officials (Millar and Len 2018). Interviews conducted across three provinces revealed a growing climate of fear driven by local CPP officials allegedly abusing their power to deprive suspected opposition supporters of essential government services (Millar and Len 2018; see Chapter 5 for this in a domestic violence case).

Given all of these developments, David Chandler, known as Cambodia's foremost historian, admitted in April 2018 that his approach to updating the fifth edition of his original 1983 work *A History of Cambodia* will need some revisioning. He reveals that, 'I just saw more possibilities for positive political changes in Cambodia ... opening up and liberalizing and so on. I saw more of them in 2005 when I wrote the last edition than I do now. I think the political scene has contracted and it's become less open to a positive change, or democratization' (cited in Chap 2018, np). Across a similar horizon scanning from the mid-2000s until today, my own book is of the same persuasion and points to the putative peace that has become a constitutive part of political and social life today. Two tropes about Cambodia perhaps best illustrate this and are particularly relevant to the empirical chapters that follow, which retell and rework grand narratives of change about the country through the extra-domestic home.

From Battlefield to Marketplace

The 'battlefield to marketplace' transition (Glassman 2010) that Cambodia has undergone has gathered pace over the course of my research. The country has been described as moving from virtual isolation to global connectedness and with this has become known for more than Angkor or the Killing Fields because of the variety of influences and developments that have shaped Cambodia in recent times (Chau-Pech Ollier and Winter 2006; Gottesman 2003). *Home SOS* proceeds to complicate any clear-cut assumptions that are made about Cambodia's supposed battlefield to marketplace transition. It also shows that while a mainstay focus of (non)-academic writing on Cambodia's transition has been on the accumulation of assets and wealth by the elite, this economic-centred viewpoint perhaps misses the larger co-option of whole lifeworlds. What will be detailed in this section is the economic and political significance of change, yet there remain important stories to tell about how these influence, and are influenced by, the intimacies of everyday life. Homes, marriages, and families are the battlefields of the marketplace that demand greater attention.

Cambodia's economy has experienced drastic changes since the late 1980s, bringing tremendous growth as well as challenges. After the gradual urban repopulation and reconstructions of the 1980s and accelerating developments nationally in the 1990s, the Cambodian government's pursuit of profit has continued. Often talked of as a success story, Cambodia's economy has grown considerably with GDP growth rates of between 7 and 7.5% from 2011 to 2016 (World Bank 2016). The garment industry is the most striking driver of the country's economic boom and the strong trend for rural–urban migration to its factory floors. In the 1980s and early 1990s the garment industry produced only basic materials for domestic consumption, such as skirts, blankets, scarves (*krama*), and some medical materials. Even this production was of low quantity. By 2005, however, garments accounted for 80.4% of total exports by value (World Bank 2006). From a handful of factories to over 600 today, 90% of its workers are women (Lawreniuk 2019). The construction sector has also played a crucial role, with Phnom Penh's literal ascent over the past decade driven by domestic and foreign capital investment in the building of elite housing, shopping, and business developments (Brickell et al 2018; Paling 2012) (Figure 3.3).

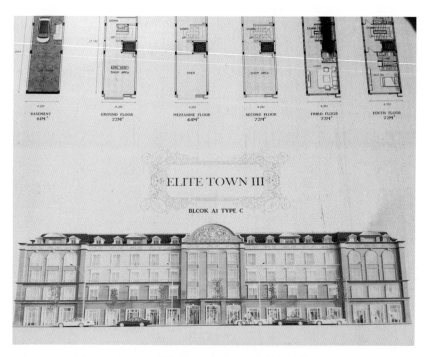

Figure 3.3   Elite Town III, Koh Pich, Phnom Penh, 2019. Photo: Katherine Brickell.

However, both these drivers, garments and construction, speak to the creation of low-paid work and poor social protection as the prevailing economic model (Cook and Pincus 2014). As I set out in Chapter 1, the majority of interviewees' households in *Home SOS* are characterised by insecure and low-paid work in farming, petty trading, and tourism. Furthermore, while Cambodia had a high labour force participation rate of 80.9% in 2017 for women over the ages of 15 (United Nations Development Programme (UNDP) 2018), women are largely confined to insecure, poorly paid positions. Women are under represented in higher-level occupations and decision-making positions in the country (Asian Development Bank (ADB) 2012). The Office of the High Commissioner for

Human Rights (OHCHR) (2015) cites that 70% of women in Cambodia are engaged in what is classed as vulnerable employment. Data from the *Cambodia Labor Force and Child Labor Survey 2012* (International Labor Organization and National Institute of Statistics, 2013) also highlights related economic inequities that women face in respect to wage work. Average monthly wages among women wage workers aged 15 years and older were 418 808 riels per month (about US$104), compared with 518 202 riels per month (about US$128) for men. Thus, women earned 80.8% of men's earnings (Asian Development Bank (ADB) 2015, p. 13). There also remain serious challenges to women's economic empowerment in Cambodia more broadly. The amount of time required to fulfil responsibilities in unpaid domestic and care work and women's low levels of literacy and education are notable inhibiting factors. In respect to the former issue, the Human Development Report (UNDP 2015, p. 125) includes time use data from 2004 in Cambodia, which indicates that women undertake 188 minutes per day of unpaid labour in comparison to 18 minutes by their male counterparts. Overall, the widespread nature of male biases in Cambodian society helps perpetuate gender inequality in terms of resource allocation and distribution, it legitimates the silencing of women, and the biases within state institutions result in structures and policies that perpetrate these inequalities. It also means that marriage still has a pronounced economic as well as social rationale for Cambodian women faced with these biased and precarious circumstances.

More generally, disparities in wealth have become more pronounced in the country as it followed its marketplace transition. Development in the agricultural sector has been slow with uneven development leaving many rural Cambodians (who make up around 85% of the total population) in poor conditions. Although the government has emphasised the need for rural development in its plans, they have offered only 'a long wish list of things to do, achieve, or accomplish', all geared to 'provide', 'to improve', 'to foster', or 'to upgrade', but with no substantial tangible outcome (Curtis 1998, p. 62). Poverty is overwhelmingly concentrated in rural areas, and the gap appears to be growing according to the ADB (2014); whereas 89% of poor households lived in rural areas in 2004, this had increased to 91% by 2011.

Elite power over land title procurements is a key example of how 'neoliberalism frequently opens opportunities for well-connected government officials to informally control market and material rewards, allowing them to easily line their own pockets' (Springer 2011b, p. 2555). Large-scale investments in land, for example, typically come about 'because of strong backing from powerful government officials and local tycoons and elites who are involved in almost all foreign joint ventures' (Young 2016, p. 598; see also Un and So 2011). The country thereby faces renewed encounters with violence arising from predatory and extractive capitalism (Springer 2009, 2010, 2015) that is fuelled from within, and beyond, Cambodia. The coexistence of commodification and privatisation together with forceful expulsion echoes what David Harvey (2003) calls the 'accumulation by dispossession' in which assets and wealth are transferred from the masses to an elite social stratum. The tenure insecurity and protest that has arisen in Cambodia must be understood both as a problem and response that is locally manifest yet implicated within a broader set of political and economic processes connected to neoliberalism (Biddulph 2011; Springer 2010). Corporate cum government power in Cambodia has aligned development with the necessity to bulldoze homes. The case study example of Boeung Kak Lake (BKL) exemplifies this, as ordinary people's homes have been sacrificed for the building of high-end residential and commercial complexes (Figure 3.4).

Figure 3.4   One Park development built on the former Boeung Kake Lake, 2018. Source: © Royal Holloway, University of London. Courtesy of Thomas Cristofoletti.

With echoes of the 'Khmer Rouge leadership [who] demanded an unmaking of space as a means of ushering in a communist society' (Tyner 2008, p. 106), elite power today demands the unmaking of home to quite literally cement Cambodia's predatory economic development through large-scale infrastructure projects. Forced evictions are therefore not only present in the country's past but also to the forefront in its present. Forced eviction and land grabbing in Cambodia have been the subject of substantial writing in academia reflecting the enormity of the issue (Arensen 2012; Beban, Sokbunthoeun, and Un 2017; Beban and Work 2014; Brickell 2014a; Connell and Grimsditch 2017; Hennings 2019; Kent 2016; Kry 2014; Lamb et al. 2017; Richardson et al. 2016; McGinn 2013; Milne

2013; Oldenburg and Neef 2014; Park 2019; Sovachana and Chambers 2019; Springer 2013a, 2013b; Schoenberger and Beban 2018; Talocci and Boano 2015, 2017). Abdou Milaq Simone (2008, p. 190) observed in Phnom Penh that 'the intensity and speed of these forced displacements have reached crisis proportions, generating almost daily coverage in the popular press and a topic of everyday conversation amongst city residents'. This crisis has not abated but has only intensified in the intervening twenty years.

Land grabbing is a process that affects the original users or owners of the land who are generally poor smallholder farmers or people using communal land. While there has been some criticism of the 'hype' that land grabbing has received in coverage by multilateral organisations, journalists, civil society organisations, and academics (Kaag and Zoomers 2014), Michelle McLinden Nuijen, Prachvuthy, and Van Westen (2014, p. 153) argue that 'the evidence is sufficient to state that land grabbing is not a hype in Cambodia' where '"development" in the guise of logging and conversion of land into large-scale agricultural operations (often rubber plantations and cattle rearing) is making significant inroads' (see also Beban et al. 2017). Almost 56% of all arable land is given to private companies for agro-industrial use in a country where 80% of Cambodia's 15 million population depends on land and natural resources for a livelihood (Association of World Council of Churches Related Development Organizations in Europe (APRODEV) 2011; see also this report for an overview of land ownership historically in Cambodia).

Forced relocation has therefore been increasing owing to a range of developments, from agro-industry to major infrastructure, hydropower, urban beautification, and private development projects (Connell and Grimsditch 2017). The entrepreneurial Cambodian state has become equivalent to 'the acid of geoeconomic calculation' (Cowen and Smith 2009, p. 41), linking citizens' homes and bodies with a broad set of political and economic neoliberal processes (Biddulph 2011; Springer 2010). Many of these forced evictions and land grabs are connected to the rise of China as a global economic and political actor. Over the past 10 years 'the Sino-Cambodian relationship has reflected a newly evolving geopolitical landscape characterised by flexible and opportunistic arrangements' (Burgos and Ear 2010, p. 615; see also Deth, Moldashev, and Bulut 2017 for an overview of the contemporary geopolitics of Cambodia). While China has become the biggest source of Foreign Direct Investment (FDI) to Cambodia, this 'evolving geopolitical landscape' links to wider changes in territorial politics since 1997. Post-financial crisis, China rescaled its centralised power structures to embrace economic development via 'transnational regionalization' (Su 2012) to Africa (Power and Mohan 2010) and South East Asia (Dwyer 2014; Glassman 2010). As part of the Greater Mekong Subregion (GMS) programme, Cambodia became the recipient of Chinese-backed aid packages and preferential loans. Unlike their Western counterparts they are not conditionally tied to democratic reform and good governance. This has created apposite conditions for the marrying of politics and business – a relationship that has resulted in well-connected elites exerting power over land title procurements and their material rewards (Springer 2011b). China's Belt and Road Initiative (BRI) is also starting to have traction in Cambodia and cements China as its premier partner. Cambodia's broader pivot to China also has a cushioning effect against Western pressures and has resulted in a decline in Western leverage (Un 2019).

Order and Stability

A second prevalent political trope, cum biopolitical tactic, noticeable in Cambodia is that of order and stability. Joel Brinkley (2011, p. 354) argues in *Cambodia's Curse* that 'For all his faults, Hun Sen has given Cambodians one very important thing: more than a decade of stability and calm that bring[s] some predictability to their lives for the first time in centuries'. Political stability, however, has come at the expense of meaningful democratic pluralism (Lawreniuk and Parsons 2018). The prerogatives of order and stability are, by extension, protective of business interests and capital, which have been put ahead of genuine democratic awakening (Springer 2009). Neoliberalisation, it is argued, suffocates democratic politics, its 'asphyxia-tion is brought to bear under the neoliberal rhetoric of order and stability which can be read though the (re)production of public space' in Cambodia (Springer 2009, p. 138). *Home SOS* shows how this suffocation is not limited to public space, but is also operative in domestic life. As the empirical material that follows substantiates, this point has profound consequences for women's lives in particular.

Order and stability are contrasted with the former Khmer Rouge regime and the prospect of opposition parties holding power. CPP 'discourse draws on memories of past conflict to re-inscribe a distinction between the ruling party – associated with stability and prosperity – and the opposition party – associated with a disorder that threatens to plunge the country back into chaos' (Schoenberger and Beban 2018, p. 1343). In doing so, the crisis ordinary is convened through the repeated evoking and enfolding of former violences into the present. Speaking at the 9th EU–Cambodia Joint Committee meeting in Phnom Penh on 4 May 2016, Ouch Borith, a Secretary of State at the Foreign Affairs Ministry, told EU diplomats (cited in Willemyns 2016, np) that the CPP had 'no choice' but to arrest its critics: 'Our unique history made "peace and stability" the core value to be preserved at all cost.' Activities deemed to 'jeopardise peace, stability, public order or harm national security, national unity, culture, customs, and traditions of the Cambodian national security' (Article 8 of LANGO) have therefore been subject to a 'legal offensive' (Subedi 2011, p. 252). This speech demonstrates how a historical period of crisis can be mobilised as 'politically expedient because it leads beyond its immediate condition, justifying policies and responses that are dispropor-tionate, or out of line or context' (Adey, Anderson, and Graham 2015, p. 6).

Cambodians have long been told to keep to their station and Hun Sen wreaks fear on the dangers of revolt as an uncivil act that may lead to an all-out crisis akin to the Khmer Rouge. These scare tactics are not new, however, and were common throughout the decades of war and after the CPP came to power in the 1980s (Hunt 2014). In this book, I demonstrate how the political emphasis on stability and order are both causal and symptomatic of the crisis ordinary that Cambodians encounter in their everyday lives. The Cambodian government view is grounded in the portrayal of human rights as antithetical to its security, rather than as a means of achieving citizens' security. As Defence Minister Tea Banh (cited in Sokhean 2016, np) portrays, 'the excessive use of [human] rights will bring about destruction, broken families and loss of hundreds of thousands of lives, loss of habitat, bloodshed'. A powerful sense of obligation is therefore cynically pressed upon Cambodian citizenry in efforts to make the rejection of human rights more widely shared.

Political discourse positions human rights as Western-imposed concepts that have a dangerous potential in a country that has, since 1993, stressed freedom from outside cultural influences that undermine 'models of behaviour that are believed to represent a true Cambodian society' (Jacobsen 2012, p. 90). As Evan Gottesman (2003, p. 12) explains, 'Cambodia has been colonized, overwhelmed by foreign ideologies, torn apart by international conflict, and occupied by other countries.' It is therefore perhaps understandable that given 'these incursions, Cambodians have attempted to define an uncontaminated national identity' (p. 12). As *Home SOS* goes on to explore, however, traditions of deference, fatalism, and hegemony have seen a growing challenge. While deviation from them invites chaos, as per CPP rhetoric, the costs of not doing so have become unbearably high when home loss so readily beckons. These losses are the central focus of the first empirical chapter, which follows looking at the SOS calls being signalled in crisis ordinary Cambodia.

Endnotes

1 Marshal Lon Nol was a Cambodian politician and general who served as prime minister of Cambodia twice (1966–1967 and 1969–1971).
2 Before French colonisation, which began in 1863, Cambodia was governed by customary rules based on consensus. From 1863 to 1953, however, the Cambodian legal and judicial systems were based almost entirely on the French system. The French Protectorate introduced the code civil – the legal framework for all civil matter rights – which included family law regulating marriage and divorce (Ebihara 1993; Heuveline and Poch 2006; Népote 1992). The French system was retained after Cambodia's declaration of independence.
3 For the full constitution see: http://www.d.dccam.org/Archives/Documents/DK_Policy/DK_Policy_DK_Constitution.htm.

4 While there is general agreement that sexual violence characterised the period more than previously thought, there exists debate about the liability of senior cadre and what this means given the ECCC's focus only on the most senior of Democratic Kampuchea leaders (see Hughes and Elander 2017 on the ECCC).

5 FUNCINPEC National United Front for an Independent, Neutral, Peaceful and Cooperative Cambodia, is a royalist political party in Cambodia. The Khmer People's National Liberation Front (KPNLF) was a political front organized in 1979 in opposition to the PRK regime. The People's Republic of Kampuchea was founded as a communist group dissatisfied with the Khmer Rouge.

Chapter Four
Attrition Warfare, Precarious Homes, and Truncated Marriages

Introduction

This chapter pursues a bio-necropolitical reading of contemporary Cambodia to explore women's narratives of their precarious homes and truncated marriages. Domestic violence and forced eviction, I argue, are both intimate wars that Cambodian women are traversing through multiple forms of death. The chapter therefore heeds the point made that any focus on deathscapes 'should encompass closeness to death and its multiple messy emotional politics to fully recognize the continuum of living/s–dying/s' (Stevenson, Kenton, and Maddrell 2016, p. 162; see also Maddrell and Sidaway 2016). As I set out in Chapter 2, this continuum is reflective of biopower and necropolitics as 'two sides of the same coin' (Braidotti 2013, p. 9), lying at the margins of life where there is overlap and difficulty in distinguishing between them (Stepputat 2014; McKinnon 2016). As Jasbir Puar (2017, p. 35) follows, this 'bio-necro collaboration' is cognisant of 'biopower's direct activity in death, while remaining bound to the optimization of life, and necropolitics' nonchalance toward death even as it seeks out killing as a primary aim'.

At this intersection of life and death, I consider women as bio-necropolitical subjects along a 'life–death continuum' in which 'affirmative practices' (Braidotti 2013, p. 130) are used to cope with violences in and of the home, which have gendered, psychological, economic, and structural inflections. As I set out in preceding chapters, domestic violence and forced eviction are intimate wars in themselves, but together they are part of a much larger war on women. The interview

Home SOS: Gender, Violence, and Survival in Crisis Ordinary Cambodia, First Edition. Katherine Brickell.
© 2020 Royal Geographical Society (with the Institute of British Geographers). Published 2020 by John Wiley & Sons Ltd.

material I engage with does not detail active public or organised opposition to these intimate wars (see Chapter 5); rather the chapter speaks to the daily activity of endurance and attempts to (re)gain recognition and respect under circumstances that might otherwise be viewed as unremarkable and unnecessary to remark upon. Women's undertaking of this survival-work to live on, through, and via mundane domestic practices of resilience is the central focus of what follows.

The chapter shows how this survival-work is performed under conditions of attrition warfare, popularly understood as a military strategy that consists of wearing down the enemy to the point of collapse. I argue that the dual study of domestic violence and forced eviction reveals a similar tactic in civilian life and has depletive effects. These are intimate wars that can be described as 'slow violence', that which 'occurs gradually, and out of sight, a violence of delayed destruction that is dispersed across time and space, an attritional violence that is typically not viewed as violence at all' (Nixon 2011, p. 2). The focus on the crisis-affected home shows how this attritional violence manifests in material and symbolic ways within marriages that are forged, and foreclosed, in the domestic realm.

The chapter is thus attentive to how women are constructed as responsible for and undertake the uneventful social reproductive work of homemaking under precarious conditions and influences of the past and into the present. As I set out in the previous chapters, women in Cambodia gain important public value through domesticity and have primary duty for the social reproduction of their households and the successes or failures of their marriages. By studying both domestic violence and forced eviction, altruism is revealed as a key behavioural register of the crisis ordinary. Familial altruism is a form of labour that has emotional as well as physical parameters and is part and parcel of the normative load that women, in particular, are expected to carry (Brickell and Chant 2010; Calkin 2015).

The significance of altruism in the repertoire of what manifests as the crisis ordinary is perhaps not surprising given that 'crisis is not exceptional to history, or consciousness but a process embedded in the ordinary that unfolds in stories about navigating what's overwhelming' (Berlant 2011, p. 10). As M.V. Lee Badgett and Nancy Folbre (1999, p. 316) explain, social norms 'assign women greater responsibility for the care of dependents, an assignment that almost literally requires altruism'. This is especially true in women's navigations of domestic violence and forced eviction. Rather than a 'natural' female attribute, 'it is important to recognise that such altruism is often a manifestation of their disempowerment, a response to their restricted options' (Kabeer 1999, p. 41). Khmer women who do not follow culturally endorsed expectations of female altruism, for example, fear adverse consequences, from divorce to other forms of social isolation, which the chapter highlights. Although it is undeniable that 'intimacy builds worlds' (Berlant 1998, p. 282), it is also the case that domestic violence and forced eviction are felt in such adverse terms precisely because of their rootedness in, and articulation through, intimate gender-based violences against women and their homes. Their collateral damage cuts deep.

Fearing further loss, the prospect or lived experience of 'social death' is a powerful vector of compliance to gender norms in Cambodia, which emphasise altruism much more strongly for women than men. Women who transgress the moral bases of everyday life tied to marriage, regardless of circumstance, are themselves embodied *as a crisis*. Penny Edwards (2008, p. 232) writes that when this happened historically in Cambodia, women were seen as violating a social code, as 'gate-keepers between the wild and the civilizing, opening society up to moral rot'. They are 'invested as icons, guardians and boundary-markers of national identity' and therefore 'women's bodies are routinely held up by state actors as embodiments of the idea of national sovereignty' (Edwards 2008, p. 232). This situation demonstrates a wider observation made by Wenona Giles and Jennifer Hyndman (2004, p. 14) that 'women are always included in the constructions of the general body of members of national and ethnic collectivities and/or citizens of the state; yet there is often a separate body of regulations (legal and/or customary) that relate to them specifically as women' (see also seminal work by Yuval-Davis (1997) on women as 'carriers of tradition'). Women face precarity when they choose, or are forced, to step outside the moral limits of civil society, the (social) death that can transpire operating as a form of punishment for women whose homes and/or marriages unravel as a consequence of domestic violence and/or forced eviction. Women's survival-work and despair heightens when there is no longer a home to protect and build life from, or when any number of altruistic and placating behaviours are not enough to halt or cope with domestic violence and the negative impacts of forced eviction.

In Chapter 1, I introduced these lines of enquiry by focusing on Orm, an interviewee whose life story spoke to the importance of studying domestic violence and forced eviction in tandem. To provide more detail on these, the chapter is divided into two main parts, the first on domestic violence and the second on forced eviction. In each I present and analyse women's experiences, which have come to characterise crisis ordinary Cambodia. By bringing domestic violence and forced eviction into analytical coalescence, I show how women encounter multiple and interconnected 'unequal regimes of living and dying' (Haritaworn, Kuntsman, and Posocco 2016, p. 3) that deplete their bodily integrity and homemaking capacities. Domestic violence and forced eviction are not discrete violences; rather they intimately coexist in the life histories and experiences of Cambodian women who are dealing with intimate war waged on multiple fronts and sites.

Domestic Violence in Cambodia

Domestic crises are typically long-running and can affect women's lives for years, even decades. They often become subsumed and normalised into the weave of everyday life such that women's bodies and homes feel like they are under almost constant threat. In this half of the chapter, the complexities of living and dying are emphasised by women who describe their lives as attuned through domestic violence. Here I focus specifically on survivors Nakry and Soportevy and the intimate wars of attrition that they have endured. I then move on to consider the altruistic politics of managing this crisis ordinary.

Intimate Wars of Attrition

Nakry was 50 years old when I met her in Siem Reap Province. A widowed cakemaker supported by her five daughters who sold soup, she had been the target of verbal and physical violence by her alcoholic husband for decades. Nakry felt physically confined and emotionally trapped in a marriage that denied her 'freedom' in spatial and cognitive terms. The couple were matched by Pol Pot cadres, a practice typical of the Khmer Rouge period (1975–1979). Marriages were 'suggested' with varying degrees of force on behalf of Angkar – the supreme yet secretive authority in Cambodia. Couples would then have to wait until a certain number of couples in the village were ready to marry for the ceremony to take place (sometimes even 100). At this time, according to Peggy Levine (2010, np), these 'group marriages, along with prescriptions for sex, pregnancies and births, thus became a central feature of the remaking of Cambodian society and contributed to the dissolution of the country's ritual practices' (such as parents arranging their children's marriages). Nakry's expressed lack of sentiment towards her deceased husband reflects Khmer Rouge attempts to dissolve the idea of romantic love and compatibility so that personal happiness would not be indulged at the expense of the work needed to reconstruct the nation. Couples, for example, were only allowed to call each other *mit p'dai* (comrade husband) and *mit bprapouan* (comrade wife) (Jacobsen 2008, p. 223), with working teams replacing family and household as the new basic socio-economic unit (Ebihara 1993). Expressions of love for family such as weeping over the death of a spouse or child were scorned upon and even punished (Hinton 2002).

Nakry's unhappiness was intensified by her husband's abusive behaviour after the regime fell. Many women believe that there has been a negative change in their husbands since the Khmer Rouge. In the first study of domestic violence in Cambodia, women described men as 'broken' and 'damaged' (Zimmerman 1995, p. 17). On a personal level, Nakry accredits her husband's alcohol addiction and resultant violence to its lasting psychological impacts. While the official end of armed conflict signals a change from war to peace at the political level, the impact of conflict on society alters it profoundly, and no political settlement can solve the social problems that this can cause (Pickup, Williams, and Sweetman 2001).

Research with survivors of the Khmer Rouge regime consistently identifies alcohol abuse as a means of dealing with the psychological trauma of their experiences and numbing the pain of what they witnessed (Brickell 2008; Van Schaak, Reicherter, and Chang 2011). Alcohol remains an unaddressed issue in Cambodia despite the fact that in 'every town, including isolated villages, one can find people with substance abuse problems, particularly related to alcohol consumption' (Lala and Straussner 2001, p. 323). It has also been shown that having a spouse who drinks alcohol increases the likelihood of intimate partner victimisation (IPV) and perpetration by 320% and 270% respectively (Ministry of Women's Affairs (MoWA) 2005). These patterns are not unusual in that wider research has shown the connections that exist between state-level dynamics and post-war microlevel behaviour on increased alcohol consumption as a means of coping with residual traumas (Lala and Straussber 2001; Moser and McIlwaine 2001), the quality of marital relations changing (El-Bushra 2000), and problems of readjustment and reintegration of men and women (Cockburn and Zarkov 2002; Dolan 2002). In fact, an exploratory study by Theresa De Langis et al. (2014) found that although some couples have remained in a relationship since the Khmer Rouge, half of them are dysfunctional and women report spousal abuse. In this sense, there exists a continuum of intimate war conveyed through older women's accounts of domestic life. As such, it is critical that the historical 'pre-making' of precarious conditions in everyday life are taken into account in analyses of the present (Lancione 2019b, p. 183).

As a by-product of Khmer Rouge marriage practices and the turmoil that the conflict inflicted on peoples' lives, for example, some women who had experienced domestic violence voiced a sense of relief on the death of their spouse, a final release from this intimate war. This outlook has been noted elsewhere, including in East Africa, where recurrent observations from women were that 'a woman is better off without a husband' and 'if he were only dead' (Silberschmidt 2005, p. 192). Nakry told me that the death of her abusive and controlling spouse had been a positive change:

> When my husband was alive I had no freedom. I could not go anywhere. When I was young I worked as a guide .... I worked at the Angkor temple and studied history. My husband made me lose that knowledge ... and I became a dull person. He always scolded me. He was violent so often and broke everything in the house. I did not love him .... Pol Pot arranged our wedding. My marriage was destroyed by my husband's rudeness. He did not know about honour. He did not think about earning. He drank alcohol but he was worse than a drunkard [tear in her eye]. For other women, when their husbands die, they feel like they are in the middle of the sea. For me, I am happy. After he died I could do what I wanted ... although I have only saltwater with rice I am happy. For other women, a husband is a golden mountain. For me, my husband was not – he drank us into poverty – and if I had lived with him much longer I would have died young.

Nakry's interview unearths ideas of death as a form of release. Along with a greater feeling of well-being, Nakry says that her husband's death has enabled her to fulfil latent ambitions and to spend more time devoted to Buddhist practice in preparation for the next incarnation.[1] In fact, Nakry believes if she had 'lived with him much longer' she would have 'died young' having been on the edge of defeat for some time. Nakry considers her husband's death as an acceptable 'trade-off'. The idea of 'trade-offs' is particularly relevant to many women's lives as they make tactical choices between different dimensions of poverty and well-being (Chant 2007; Kabeer 1997). Although being without a male partner is often associated with heightened levels of deprivation (only having 'saltwater with rice' rather than having an ideal husband who is a 'golden mountain' providing for his family), this can simultaneously be met by spin-off benefits as spousal violence rarely occurs in isolation from other controlling behaviours (National Institute of Statistics, National Institute of Public Health [Cambodia] 2006, p. 289). As Nakry alludes, these advantages include pursuing a more vivacious life free from violence and a spouse literally drinking the household budget. That her husband 'did not think about earning' reflects the dual burdens of productive and reproductive labour than Orm mentioned in Chapter 1.

My interview with Sopotevy also evidences the attritional cycle of violence commonly encountered. Speaking at the age of 51, she explained that she was customarily married to Thom in 1981, had five children, and made a modest income by selling vegetables outside their home in Pursat Province.[2] Soportevy dealt with regular instances of physical violence, including once being hit on the head with a metal torch. The couple separated in 1997, at which point Thom began drinking heavily. He turned to the monkhood until 2011 when Soportevy agreed to get back together with him on the understanding that his Buddhist devotion must have 'changed him'. She went on to explain what happened since:

> The situation has become so redundant that he [the village head] can only throw his hands up in the air in disgust because there isn't much he can do. Thom has been drunk for so long, sister. The only reason I got back with him was because he had been a monk and I thought that he would change. But he didn't. It had been easier just living alone with my children and grandchildren. I didn't have to worry about anything. With him, I only experience emotional hardships and it is too much. If I don't go out and find work, we won't have food to eat. He does not care about me. At this age, while I still can, I will work until I am too old for it. Then, I will try to depend on my children to take care of me for a change. My husband, with his drunken way, will never be dependable. I barely sleep at night. He usually doesn't come home until one or two in the morning, completely drunk. He yells and causes so much commotion that even the neighbours cannot sleep. You see, if he only did this once in a while, they probably wouldn't mind at all. The thing is, he annoys the neighbours with his drunken rants every single night. I have no doubt that they have very low opinions of my family. They look down on him – and me – because he is so bad. He has no understanding of even the most common sense courtesy around here …. I just know that this is going to be like this until the day that I die. I also know that he will never change, regardless of who talks to him. I just consider this as my fate, something that I must bear. If I happen to die before he does, then it will be my blessing to go earlier. If he happens to die before me, then I will rejoice with the time I have left without him.

As Soportevy's interview illustrates, the domestic violence she encounters has been slow in the sense of its attritional potency across her life course, from the past into the present, and stymieing her imagined future. This intimate war of attrition is perpetrated through a range of tactics. The first relates to the divergent yet concurrent temporalities of violence that paralyse Soportevy's life. Thom not only causes problems 'every single day', but these are repeated over time to the extent that 'he will *never* change' (my emphasis). Her husband's alcohol dependency feeds this sense of her situation being intractable. The second tactic is through sleep deprivation and poor quality of sleep which is a major issue for domestic violence victims like Soportevy (see, for example, Woods, Kozachik, and Hall 2010). Third is the sense of hopelessness, which has congealed over time given the redundant option of help-seeking after multiple failed attempts (see Chapters 5 and 6 for elaboration on these failures to support women's external help-seeking). Expressed hopes, for example, that her husband would change having entered the monkhood were dashed and, much like Orm, hopelessness had taken an affective hold over her life. If 'crisis temporalities determine, and are determined by, how individuals experience time and space under states of emergency, assigning meaning to these durations and places' (Itagaki 2013, p. 197), then for many women domestic violence survivors, the home and their everyday encounters inside and outside of it are experiential spaces and registers in which crisis is not limited to a discrete site or period. In fact, there seems to be few definitive ends to the attenuation of life, apart from death.

The lack of options to address the seemingly intractable violence Soportevy faces is also referenced in relation to fate. While Buddhism represents an ethos of non-violence, it also 'promulgates the rather merciless law of karma, according to which your present life situation is the cumulative result of deeds in your previous incarnations' (Ovesen, Trankell, and Öjendal 1996, p. 77). Therefore, if one has a failing marriage, it is because of one's karma. Later in the interview, for example, Soportevy reiterated, 'I am looking forward to die early so that I don't have to face him any more'. In the interview, death is presented as agency, be this her imagined death or that of her husband, which would provide release from the unrelenting drunkenness and violence. The sense of death as agency is also a prevalent discourse and practice for some forced eviction victims whose experiences I turn to later in this chapter.

In recent work on the attempts of migrants and refugees to cross into Europe via sea, Kim Rygiel (2016) has written on 'dying to live'. In the broader context of work on necropolitics and ungrievable lives, Rygiel argues that 'death disrupts the logics of citizenship politics' and gives the example of death of the migrant/refugee who 'forces the state to forgo its categorization of legality/illegality in order to administer to the dead body' (p. 549). In a different but nonetheless related way, the death of either the domestic violence victim or perpetrator provides a moment of structural transgression in which the possibility of creating a new life free from violence is furnished – be this in the present world or in the afterlife. For some victims, the prospect of death feels, again, like the closest they will get to claiming rights to a life free from violence in Cambodia.

This is because women do not see divorce as an option and feel coerced into continued union at whatever personal cost (Brickell 2012c, 2014b). As a result, they undergo continuous violences and losses consistent with attrition warfare. A Siem Reap-based interviewee explained that as a divorced woman:

> … there will be a lot of problems. Once you are divorced, it will not be easy to have any interactions with other men in the village without being considered a slut. People will be slow to sympathise with our lives. Without our husbands around, we will be looked down upon.

In cases where women wish to end a violent marriage, they can face a range of barriers to expressing agency over their conjugal lives. Briefing information from a Legal Aid of Cambodia workshop on access to divorce portrays the likely discourse: 'Had enough of an abusive husband? Get a divorce! So goes conventional wisdom, but for many Cambodian women, the challenges and social stigma attached to divorce means they would often prefer to suffer in silence than go through the process' (United Nations Development Programme (UNDP) 2010, np). Censure from the community in Cambodia acts 'as an extremely strong deterrent to divorce, particularly for women … a woman is marked for life, as a disgrace to her family, as an unfit marriage partner, as "used goods"' (Ledgerwood 1990, p. 181; see also Derks 2008).

Interviewees feared what is tantamount to 'social death' should they divorce given that 'a serious loss of social vitality is a loss of identity and consequently a serious loss of meaning for one's existence' (Short 2016, p. 34). This prospect of social death has echoes with 'the gendered punishment of living death', which Christen Smith (2016, p. 31) tells of in her work on the gendered necropolitics of anti-Black state violence in Brazil and the United States. She argues that:

> Although state terror often results in the immediate physical death of young Black men, it is principally, yet tacitly, performed for Black women and impacts Black women disproportionately. Black mothers (social, biological, or otherwise) are scripted within the racial, hetero-patriarchal social order as enemies of the state. As such, they pose a unique political threat to the social order.

Cambodian domestic violence victims who dare to defy norms of familial harmony and togetherness to divorce are stepping outside the social and moral order that Cambodian tradition and government rhetoric reinforces. They are a 'political threat', a source of potential political crisis, which cannot be countenanced. That the importance of familial harmony has been legislated into Cambodia's domestic violence law (DV law) further supports this point (see Chapter 5 for an analysis).

Altruistic Politics of the Crisis Ordinary

In this second section on how women negotiate the intimate war of domestic violence, the stories of Chantrea, Kesor, Darareaksmey, and Kalliyan are used to show how women's lives are being further truncated to protect those of their (ex-) husbands and children in a political power play intermeshed in patriarchal, state, and community norms. While these apparent self-penalising acts make many women feel like the 'living dead', they are nevertheless crucial to women's ability to avoid social death. It remains women's normative role in contemporary Cambodia, and in capitalism more generally, to subordinate their needs to ensure those of the familial collective whose lives are understood to be more worthy and therefore more grievable. In this section, women's narratives testify to the altruistic necropolitics of quotidian life that are deemed necessary to ensure the social reproduction of the household and, by extension, the nation.

Chantrea was 49 years old when interviewed and had been married informally twice; her first marriage was after the fall of the Khmer Rouge in 1979 and her second in 1983 to a truck driver. Her interview demonstrates the perceived impunity that protects her second husband from any action taken against him. Akin to Orm and Soportevy, a sense of hopelessness and despair pervaded her life story:

> My husband hits me regularly and even told me to report him to the village head. But in my heart I thought that I couldn't do this because now and later he is my husband. What if I went and report him, surely he would then kill me? He hit me so hard while I was still eating that I passed out with the rice still in my mouth. That was the first time. The second time he beat me to the point that I bled from my nose .... I should have saved all those bloodstained cotton swabs to show a human rights group ... but I am afraid he will harm me ... he told me I could do whatever I wanted, report him, but he'd hurt me even more for it ... honestly, I just want to cry ... the blood was gushing out like water. I yelled to a neighbour 'please help me' but no one dared to help.

In raw and direct terms, Chantrea cites fear of being murdered as a major factor that explains her continued marriage and ensuing isolation. With eight children and two grandchildren to also support, seeking help from a human rights group was deemed by Chantrea to be a precarious move and likely futile if her neighbours' lack of intervention was anything to go by (see Chapter 5 for further discussion on help-seeking). While my later analysis of forced eviction narratives from Boeung Kak Lake (BKL) show how such perceived futility was ignored and defied, the commonplace calculation amongst nearly all the domestic violence victims interviewed was that communicating fire outside of the house via help-seeking was likely to be pointless.

For example, Kesor married her husband Kiry in what she called a 'poor man's marriage' (a customary ceremony) in 1989. She met him through cousins who introduced them. In 2008, her taxi-driving husband insisted that she sell all her family jewellery to buy a car. Having done so, she subsequently discovered that most of the money earmarked for the car purchase had been used to pay for sex workers and other girlfriends. One year later, after the birth of their first child, Kiry begged her for more money to buy a car. Kesor told him in no uncertain terms that she would not agree given he had been unfaithful. Kiry proceeded to organise a formal land title for her family's land. She followed Kiry's instructions to press her thumbprint on to the documentation.[3] Without her knowledge, he then took out a high-interest loan from ACLEDA bank using the land value (US$10 000) as collateral to buy the car.[4] He then drove away.

Kesor described crying so much at his departure that she felt she was going blind. Debt is a growing part of the intimate war on familial life in Cambodia, with the proliferation of a deregulated credit industry from the early 2000s onwards taking its toll. This toll of debt and material hardship is endured not only by domestic violence victims whose husbands deceive them, but also in cases of forced eviction, which I discuss later.[5] Since 2008 Kesor has struggled to provide enough food for her three children. She makes minimal returns from selling fish and often takes her children to her mother's house to be fed. In 2011 Kiry returned to the village having spent the intervening years living with a 'factory girl' (a garment factory worker). Given the financial destitution faced, Kesor got back together with him. This was despite her recognition that 'I know this is considered a type of violence because he holds me in his grip when it comes to money'. Her marriage continued to be an unhappy and increasingly violent one as Kiry regularly raped her when drunk; alcohol was once again present in this intimate war.[6]

Despite the unceasing violence, and whilst angry, Kesor cited a range of reasons for her continued marriage to Kiry. Two excerpts from her interview reveal these:

> I don't like the fact that my husband treats me badly because he thinks he is pure gold and can do no wrong .... I thought about divorcing him, however one of my cousins came and talked to me. She was a child of a divorced family and urged me to hold on to marriage because separation was never a good idea. She told me to think from a child's perspective. And since then, I have stopped entertaining the idea of separating. But it's just too much sometimes, sister. I always have girls calling him late at night....

> Maybe outwardly villagers wouldn't despise me [if I reported the marital rape], but they will talk behind your backs, that your family loves violence. Even the children will suffer. They become embarrassed among their friends. Seeing their friends growing up in good households makes them ashamed of their own situations. They will be shunned at school too .... I immediately thought of how other people would view him negatively. They would look down on my husband. I wanted the problem to stay within the family .... I never want to see my husband's reputation tarnished in the community.

Throughout my time researching in Cambodia, men and women I have spoken to agree that although men may create problems within households, they remain akin to 'gold'. When gold is dropped in mud, it does not lose its value, unlike 'cloth' (women), which is irreparably soiled. This proverb remains a frequent source of metaphor used to authenticate the differential statuses of men and women (Brickell 2007) and is one that Kesor draws on directly. Kesor has received advice and is cognisant that public knowledge of the violence she experiences has its own set of risks. While on the face of it the concern for her husband's reputation might seem perverse or counterintuitive, Kesor's status and that of her children are calibrated against Kiry's. The centrality of marriage to Khmer women can also be seen by returning to Soportevy's interview:

> Look here, sister. This is how much I love him. No matter what he does to me, whether beating me and leaving me with bruises and scars, I would lie to people that I got dizzy and fell. I would not tell them the truth because I fear that his reputation would be tarnished. As a man, one's reputation in the community is very important and I want to make sure that he receives the best treatment from other people even if I have to suffer for it. I take good care of him. When he goes out to drink, I still make sure that I reserve food for him and tell my children not to touch their father's food. I'd rather save it for him so that when he comes back, he can eat. If he doesn't, I would let the food go bad and throw it away before I let my kids touch it. This is how I treat my husband.

Soportevy's narrative speaks to how women's societal statuses remain judged against intact homes and harmonious marital relationships, which are undermined by men's violence against women in cases of domestic violence (and as I explore later too, forced eviction). While she admits seeking advice from a local leader about her husband's drinking, she has been reticent to speak publicly about the violence that she has endured over the course of her relationship (see Chapter 6 on the heightened risks that women can face when help-seeking). As she notes, excuses were made for why she might have bruises or scars. The Ministry of Women's Affairs (MoWA 2009, p. 7) notes that in Cambodia '[a]cceptance, toleration, and rationalization of such abuse is deeply embedded in traditional socio-cultural value systems'. Under such social norms, 'it is indeed possible that altruistic behaviours stem not from any positive feelings of the individuals involved but, rather, from a coercive social milieu' (Vanwey 2004, p. 740).

In Cambodia, wives are tasked with the care of children, responsibility for the household economy, and ultimately with supporting the accomplishments of their husbands (Surtees 2003). As Peg Levine (2010, p. 32) explains, in Cambodia 'love within marriage is often seen as a bonus, while family harmony is the cake of tradition'. Cambodian women, including Kesor and Soportevy (and forced eviction survivors later), see themselves as squarely responsible for maintaining this 'cake of tradition' in order to protect themselves and their children from societal stigma and shame. Mona Lilja (2014, p. 31) writes that in 'present-day Cambodia there is a pattern of keeping the "traditional" gender imagery alive'. This returns to, and legitimates still further, fears of 'social death' voiced by women should they divorce.[7]

The reluctance to divorce was acute in Darareaksmey's case, a 40-year-old woman who became legally married to Munny at the age of 20. Living with her parents initially, after the birth of their first child the couple moved into a separate house (in total, they had two daughters and two sons). Relations between the couple quickly deteriorated and Darareaksmey fled to her sister's home on a frequent basis to escape Munny's heavy drinking and repeated bouts of physical violence towards her.[8] Each time she reassured herself that as Munny mellowed with age he would stop hitting her, but his 'bad habits' did not abate. Desperate, she went to the police but was asked for US$10 to initiate divorce proceedings. Unable to pay this, she was advised to contact a Non-Governmental Organisation (NGO) that helps women in crisis. Having made contact, the NGO filed for a divorce, which was granted in 2004. With traditional rural houses designed to be portable, soon after Munny dismantled their wooden house, loaded it on to a car, and left.[9] Darareaksmey explained that:

> When my husband took the house I just felt indifferent, maybe I just couldn't deal with any more bad karma between us. When we were together it was so very difficult. You know, when I went to ask someone for help at the police station, they wanted to arrest him, to sentence him, because he hit my head so hard it almost shattered, but I said that no, I needed to get a divorce through a more peaceful method … now I can sleep comfortably in the day and night whereas for the years we were together it was like living in a prison. You can ask my daughter, she never got enough sleep at night because of his drinking nearly every day. Sometimes he threw the pots around. At the organisation they asked me why I continued to live with him so long, having so many children, and I said that no one woman wants to get a divorce from her husband. I did not want to live as a widow woman.

Darareaksmey emphasises the omnipotent importance of marriage for maintaining private and public dignity. Otherwise it is ventured that she might become a 'widow woman'; a commonplace parlance and necropolitical subjectivity used by, and about, Cambodian women who are separated or divorced from their husbands. In accepting the necessity to end her marriage, Darareaksmey emphasises the trade-off she was resigned to make, between material disadvantage on the one hand and mental health on the other. In Cambodia, after divorce most ex-husbands rarely send money to support their children (Hagood Lee 2006), a trend that was exacerbated by the personal circumstances of Darareaksmey and Munny who are no longer in contact.[10] Darareaksmey has also been forced to borrow US$1200 from her relatives and other lenders to rebuild the house and purchase some rice fields. However, fearful of marrying a man of similar disrepute, opting for *de jure* female headship has been an important strategy for Darareaksmey, allowing her to finally sleep comfortably and end the cycle of domestic violence she suffered.[11] As Sylvia Chant (2007, p. 37) reflects through her comparative work in Asia, Africa, and Latin America:

On the surface, female household headship may appear to be detrimental to the well-being of women and children, especially given that women often lose out materially from divorce or separation, and particularly when they themselves take the decision to initiate a split … empirical evidence suggests that some women's desire to exit negative relationships is such that they are prepared to make substantial financial sacrifices in order to do so.

Despite mounting debts, dissolving the marriage on the most 'peaceful' terms was of the upmost significance to Darareaksmey so that no more 'bad karma' or disgrace would result. Karmic status is a particularly important concept in Cambodian society, representing the cumulative merit of deeds in men and women's previous lives, and rewarding women's proper behaviour in the present through a lifelong marriage (Ledgerwood 1994). Cecile Jackson (2012, p. 7) explains on this front that 'transitions between the symbolically complete and the incomplete are not only effected through marriage, but this reminds us that marriage and reproduction have symbolic entailments which constitute different kinds of persons, and a failed marriage brings with it symbolic costs and challenges'. The home too has material and symbolic 'entailments', which have their own costs and challenges in situations of domestic violence, and, indeed, forced eviction. Given the outlined considerations, Darareaksmey claims she felt emotionally 'indifferent' towards her husband's seizure of the house. This could be linked to her description of domestic life prior to their separation as a 'prison' (and thus consistent with much feminist writing on the possibility of the home being a cage). In this case, the commandeering of Darareaksmey's house could be viewed as a processual moment that involved an emotionally relieving, yet financially damaging, dismantling of her relationship with Munny. Truncated homes are not just embodied in the experiences of its inhabitants then; they are also manifest in their material divestment.

Domestic Violence as Forced Eviction

As the experiences of Darareaksmey testify, domestic violence can be framed as a forced eviction from home. Both domestic violence and forced eviction are forms of displacement that have logistical as well as existential dynamics of feeling displaced. For several women I interviewed, the short and longer-term damage that domestic violence had was not just corporeally lived through physical and emotional scarring but was also known through the material harm done to the architecture of the home itself. This scarring and destruction of home in its physical form is commonplace in the forced eviction process and its outcomes.

This was the case for Kalliyan, who experienced an arranged legally registered marriage in 2005, at the age of 23, to Nhean, a man unknown to her from a neighbouring village. Trusting her parents' advice on the basis of their own successful marriage, she agreed (though feeling she had no choice), reconciling what she describes as a loveless marriage with what she envisaged as a higher priority – creating a 'good life' for their future children. These hopes did not transpire. For a long period of their subsequent marriage Kalliyan struggled with her husband's alcoholism, unexplained absences, and emotional abuse. Nhean left without warning twice before returning, the first time when she was pregnant with their first child and the second time immediately after giving birth to their next child. Kalliyan's vulnerability to victimization and moral isolation became heightened in 2009. When returning after a bout of heavy drinking, Nhean announced he was finally leaving the commune and never returning. With him he took their eldest son (of two boys) along with their marital home.[12] Nhean dismantled the wooden planks of the house, loaded them on to a small truck, and drove away.[13] Both forms of violence – war and domestic violence – are marked, Pain (2015) argues, by displacement of people from their homes (see also Bowstead 2015, 2017). Kalliyan's (and Darareaksmey's) predicament lends further weight to this observation.

After a couple of months Kalliyan went to Kratie Province (in Eastern Cambodia) to claim her son. She has since lost all contact with Nhean. Sitting looking over the empty field as we talked together she told me about the strategic positioning of her new home close to her parents' (Figure 4.1). Prior to that, her conjugal home was located 30 metres away and today the land is used to

Figure 4.1   Kalliyan's new house (right) next to her parents' (left), Siem Reap Province, 2011. Photo: Katherine Brickell.

grow vegetables. This close proximity is traditional in Cambodian society, where matrilocal residency has been preferred and the new husband comes to live in the bride's natal village (see Ledgerwood 1995 for more information on the history of Khmer kinship).

Encouraged by her parents, Kalliyan moved to symbolize renewed membership of a family unit and to ease the hallucinations she suffered at night. Caused by anxiety over a perceived increased risk of rape, Kalliyan explained that it was stories from listening to the radio and watching television that had fuelled her fear. Indeed, according to Amnesty International (2010) not only do a growing number of rape reports fill the Cambodian media, but most police, NGO workers, and public officials working with the issue agree that rape is on the increase. Whilst the significance of Kalliyan's newly positioned house should therefore not be understated, the everyday security it afforded was only partial. Fear and hallucinations are all part of the existential uncertainty that Kalliyan feels, not only about the potential sexual violence she might encounter given her diminished status, but also because of the material depletion of her home as a secure and safe space.

Hers is a destabilising anxiety that enrolled her in crisis management. In questioning Kalliyan about why the house was built so close to her parents' she explained that the walls were not strong (being made from tree leaves) and it had numerous large gaps in the sides, which meant she is still often scared to sleep alone. Research in Durban, South Africa, by Paula Meth (2003, p. 323) also highlights the vulnerability of living in such material circumstances: 'not those of a formal home with formalized properties such as walls, doors, locks and roofs'. She points to women's experiences of domestic violence whilst living homeless or in insecure accommodation to show how 'the variable materiality of the home needs to be addressed more fully in terms of its capacity to shape and mediate experiences of domestic violence' (and in Kalliyan's example, fear of rape) (p. 324).

Second, legal precarity haunts Kalliyan. With Nhean declaring he was leaving forever, he convinced Kalliyan that a formal divorce was not necessary, that there was no need to go to court, despite her own wishes. She had initially sought help from commune authorities, but when her case was finally referred to the provincial law courts, Nhean had already left. Unsure about the legalities of her situation under the Civil Code (Royal Government of Cambodia (RGC) 2007a), concerned that she had no money to pay for a case, and intending to never marry again, she did not pursue a formal divorce and thus finds her status and rights ambiguously defined. In respect to land tenure, it is customary in Cambodia that women have recognised rights to inheritance and individual land ownership when married. Nhean moved, as per tradition, to Kalliyan's family plot when they married. With systematic land titling however, the land was jointly registered between them, leaving Kalliyan in an uncertain tenure situation should her estranged husband return to claim his ownership share.

As a lawyer who worked for Legal Aid of Cambodia and on the GIZ's project on Access to Justice for Women offers by way of example:

> ... a couple has land obtained by the wife through succession, but titled in the name of both the husband and wife. The land should have been registered in the name of the wife only, in line with the statutory system of marital property. However, now both parties have deviated from the statutory system. Should the husband be considered joint owner of the property? And if so, does the wife have a claim to unjust enrichment of her husband? Or could the husband state that the joint registration in the cadastre is an agreement to deviate from the statutory system of marital property? There is no clear answer (Van der Keur 2014, p. 12).

Despite the onus placed by numerous NGOs and international organizations on the protective function of joint land titling for married women, under the statutory system of marital property in the Civil Code (2007), marital property includes any land obtained during the marriage, except land obtained through gift or succession or bought with private property. Under such circumstances, from a legal viewpoint, even if the land is registered in the name of one spouse only, the other spouse is legally a joint owner through the statutory system of marital property (Articles 972–973). Therefore, while a wife is joint owner of the land through the statutory system of marital property (even if she is not registered as married), Kalliyan's reversed circumstances are manifest in a legal limbo-land.

Such complexities of post-marital tenure arrangements are not unusual. In October 2008, local and international media reported from authorities in a farming community in Prey Veng Province that a Cambodian couple hoped to avoid the country's convoluted and expensive divorce process and had taken radical steps after countless arguments, accusations of extramarital affairs, and charges made that the wife was not adequately caring for the ill health of her husband (as a selection, and for a photograph, see Bell 2008; Sheers 2008; and Titthara 2008). To separate their lives the husband literally sawed the 6 × 7.5 metre wooden house into equal halves, carrying his share of the property to his parent's plot of land.[14] As Dorine Van der Keur (2014, p. 12) notes, the 'right of divorced spouses to half the marital property seems to be well known in Cambodia' (Articles 972–973) with further public awareness raised via this case. The couple also divided their land into four parts, two for their children and two for them. Married for over 40 years but not legally divorced, this extreme, yet unbinding, settlement was discouraged by local authorities. Parodied as 'Divorce Cambodian Style' the case was presented as an oddity in the media.[15] Yet this instance of 'informal' divorce highlights the wider seriousness of marital breakdown in Cambodia and the lack of stall put in the legal process.

For many women I interviewed, experiences and consequences of domestic violence were both narrated through, and heavily mediated by, domestic materiality. Once held up in academic writing as an unduly overlooked 'third party in the marriage' itself (Anthony 1997, p. 2), the significance of domestic materiality has been told through Darareaksmey and Kalliyan's stories. It also gains heightened significance in the second half of this chapter on forced eviction. While in the media-reported example there was a degree of consent to carve the house in half, the other examples in this section on the dismantling of women's marital homes show a distinct lack of consent. Their removal is an involuntary one that has echoes with the processes of forced eviction that I attend to next.

Forced Eviction in Cambodia

In addition to attrition warfare detailed in relation to domestic violence, open annihilatory battle has left the Cambodian home smouldering from forced eviction (as Figure 1.5 previously showed along the Siem Reap River). In this second half of the chapter, I turn to the deathscape of violence and dispossession that engulfed Boeung Kak Lake (BKL) in Phnom Penh and examine lived realities as told through participants. While the 'house cut in half' represents the truncation of marriages arising from marital breakdown, here I explore the truncation of domestic walls and a community impacted by forced eviction. I focus on forced eviction (like domestic violence) as an intimate war of attrition and consider the significance of women's survival-work and extreme social reproductive burdens within this intimate battle. The second part of my analysis then explores what the depletive impacts of forced eviction mean for its gendered victims and the crisis temporalities of the living/s–dying/s, which extend far beyond eviction as an event.

Intimate Wars of Attrition

Located in the northern heart of Phnom Penh, BKL and the skirting districts of Daun Penh and Toul Kork originally attracted residents in the aftermath of the Khmer Rouge regime in the 1980s and 1990s. It became home to an estimated 20 000 people (around 4000 families) (Bugalski and Pred 2009). BKL was, as the Cambodian Center for Human Rights (CCHR) (2011, np) described, 'an idyllic body of water in the heart of Phnom Penh surrounded by palm trees, guesthouses, a mosque and several thriving villages. It was one of the capital's most prominent landmarks, helping to characterize Phnom Penh as a languid, tropical city of rural charm'. Yet necrocapitalism in Cambodia – 'practices of organizational accumulation that involve violence, dispossession, and death' (Banerjee 2008, p. 1543) – discounted BKL and has treated it as if it 'were uninhabited by the living, the unborn, and the animate deceased' (Nixon 2011, p. 17) (Figure 4.2).

Figure 4.2   The shattered brick line of an evicted house in BKL swallowed up by sand, 2011. Source: © Ben Woods. Reproduced with permission.

Without public consultation, in February 2007, 133 hectares of Boeung Kak Lake and the surrounding area were leased for 99 years from the Municipality of Phnom Penh to Shukaku Inc.[16] The futuristic city plans for BKL were a joint venture between Shukaku and the Chinese firm Erdos Hongjun Investment Corporation. Setting these visions in motion in August 2008, the lake started to be filled with sand, displacing water and flooding homes. Many homes became irretrievable to those who once inhabited them: literally and metaphorically suffocated by sand and submerged to death (Figures 4.3 and 4.4).

Figure 4.3   Evidence of inundated homes, 2013. Photo: Katherine Brickell.

Figure 4.4   Interviewee's photograph of their drowned home, 2013. Photo: Katherine Brickell.

Much like the smouldering landscape left along the Siem Reap river in Orm's case (see Chapter 1), BKL had become 'a desolate, apocalyptic landscape of sand, rubble, bulldozers and broken homes' (Cambodian Center for Human Rights 2011, np). Indeed, interviewees would commonly bring sun-bleached 'before' and 'after' photographs to evidence the inundation of their homes (Figures 4.3 and 4.4). Echoing mantra of indigenous groups around the world (Curthoys 2008; Wolfe 2006), 'no land, no life' and 'no house, no life' have become prevailing discourses of BKL (displaced) residents. These are not just contests over land, but over life itself; the necropolitical dimensions of disavowed socialities and kinning from land being brought into stark relief. Two interviewees explained that:

> I can say that the Chinese government murders Cambodians. I do not agree that China gives financial aid without debt return. In fact, we can see that most Chinese companies take over mines and forests. They are stripping our natural resources! Those Chinese businesses evict citizens and the Cambodian government is assisting them to do so!

> Chinese investment threatens my life. Furthermore, some people died. There were two people; one was male and one female, [both] died. They committed suicide because of unbearable pain. Some infants could not see the sunlight and died in the womb. This was caused by a public authority that protects the Chinese and Cambodian company and corrupted governmental officers, who hit citizens. This kills infants. Therefore, the Chinese government is a company that brings death. The company makes us live miserably and some died.

BKL women paint a landscape of forced eviction in which their lives are threatened and targeted for death via Chinese practices of accumulation by dispossession. Their depiction links strongly with the argument which Silvia Federici makes, that women have become a main target for persecution because they pose a threat to the developing capitalist order. As a mainstay feature of BKL narratives, the women refer to those affected specifically as 'citizens', thus invoking their citizenship rights to further the counter-hegemonic interests I explore in Chapter 5. The charge made here, and trend noted in Chapter 3, is that China is an ever-growing influence in the necrocapitalist development of Cambodia and is a threat to the country's citizens. Yet China has continued to promote itself as a 'peaceful giant' in Cambodia (Burgos and Ear 2013, p. 101). It has taken the lead in developing infrastructure such as roads, bridges, and public buildings with a clear focus on the exploitation of natural resources, furthering business ties, and gaining political advantage (Var 2016). The granting of economic land concessions (ELCs) to China, it is inferred, renders it a fallacy that the latter provides the former with aid without preconditions. While the West and Japan have traditionally provided aid as tied to good governance, respect for human rights, environmental protection, military and civil service reforms, Hun Sen praises China for its supposed non-interference. This 'no strings attached' policy essentially means that China does not require documented proof of the appropriate use of funds; nor does it insist on conditions being placed on these rights and reforms (Var 2016).

From the mines and forests of the ELCs, to tourist-focused Siem Reap, to rapidly urbanizing Phnom Penh, experiences of forced evictions are all too common, and are an ominous part of domestic life across Cambodia. As the prior chapters introduced, forced displacement has also been a feature of Cambodian history across four decades, having a prescient and analogous recent history in many women's lives (including Orm and Nakry). Amnesty International (2011, p. 7) makes reference to BKL specifically as 'the largest forced eviction since the Khmer Rouge' and note that the city as a whole has seen around 10% of its population evicted between 1990 and 2011 (p. 60).[17] The Save Boeung Kak campaign website notes an article by *The Guardian* in March 2011 as the first international reporting of the case.[18] It too frames the intimate war of forced eviction against BKL as a long-standing one:

> It was early in the morning when housewife Ngin Savoeun woke to cries for help from her neighbours. A survivor of Cambodia's Khmer Rouge regime – whose soldiers had murdered her husband in 1979 – she had not imagined she would hear such cries again in her lifetime. Yet on that night in November, she heard the screams of neighbours as they rushed from their homes around the shore of Boeung Kak Lake, located in the heart of the country's capital, Phnom Penh. And then she had to flee as well. This time, however, her home was not under threat from Khmer Rouge guerrillas, but was instead demolished by armed construction workers, hired by a land development corporation to carry out one of the capital's most ambitious new property developments (Gorvett 2011, np).

Ties between BKL residents' experiences of forced eviction and the Khmer Rouge period were also frequently mobilised in the interviews. Yet Sebastian Strangio (2014, p. 159) observes that city officials often refer to the urban poor and their settlements with the Khmer word *anatepadei*, meaning 'anarchy'. The BKL area specifically had also, by the early 2000s, gained a reputation for cheap tourist accommodation, weak law enforcement and easily available drugs (Stubbs, F. 2002), which may have contributed to this labelling. Such denigrative labelling was also apparent to some participants, one of whom vented, 'we have lived here for 20–30 years but are told we are living in anarchy, that we are squatters!' Informal systems of land claim in BKL had emerged from the 1980s onwards through continuous residence and/or use (Bugalski and Pred 2009). This included 'symbolic items of ownership, such as house numbers, minor infrastructure improvements and witnessing by local government officials of land sale contracts' (Fitzpatrick 2015, p. 76).

BKL women, fictitiously referred to as Da and Kravann, substantiated the comparison between the Khmer Rouge and their situation today:

> The authority starts to consider citizens as their rivals, takes advantage and abuses us. In Khmer Rouge times, they used guns, knives, and other weapons to kill citizens. Currently, the authority uses the label of development to corrupt, abuse, and to kill citizens. Why do I think they are killing citizens? It is obvious that the authority

is killing our mental health, household income, happiness, and future ... we die physically, financially, and mentally. Nowadays, the authority only evicts the poor ... that way they can take the poor outside of the city where they can hide poverty from other countries. This is what is meant by poverty eradication! (Da).

During the Pol Pot Regime, the thing is that citizens can be killed, but everyone is equal. Nowadays, we are not equal, and the poor will die because of starvation and eviction; for example, we face with the situation of losing our homes at any time. After, we have to relocate to rural areas where there are no employment opportunities or food to eat. We will die, for sure. Although we were killed by the Pol Pot Regime, we were provided with porridge to eat, but now, we don't have rice or porridge to eat (Kravann).

Akin to the women who had experienced domestic violence in the first half of the chapter, Da and Kravann see forced eviction as threatening and targeting their lives for violence and killing in multiple senses: material, somatic, emotional, psychological, and social. Sadly, Da recalls the barbarism that has accompanied women's peaceful activism and has led to multiple miscarriages (see Chapter 5 for full information). Legitimating their position again as Cambodian citizens, and raising a number of analogies, but also discrepancies, between Democratic Kampuchea and today, both women explain why the former could be considered *more* equitable than the current one in terms of peoples' killable status.[19]

In the academic realm, the links made between genocide and forced eviction are not new, albeit made primarily in relation to indigenous belonging and settler colonialism (Hinton, Woolford, and Benvenuto 2014; Huseman and Short 2012; MacDonald and Logan 2016). Damien Short (2016, p. 36, emphasis in original) argues, for example, that 'when indigenous peoples, who have a physical, cultural and spiritual connection to their land, are *forcibly* dispossessed and estranged from their lands they invariably experience "social death" and thus genocide'. Under this guise, genocide can be understood as the destruction of social figuration and is not restricted to direct physical killing (Lemkin 1944). For instance, methods and techniques of genocide range from deprivation of livelihoods, starvation, the separation of families, and the destruction of culturally important centres such as homes (McDonnell and Moses 2005). BKL is a 'displacement atrocity' (Basso 2016), which women frame in very similar terms to genocide.[20] This is articulated in three further ways.

First, the women highlight starvation as an expression of necropower, which was adopted during the Khmer Rouge. The deliberate exposure of the population to mass starvation was designed and implemented in the administrative practices of this period (Tyner and Rice 2016). Yet it is argued to be worse today as citizens are not provided with even the rice or porridge offered (albeit in meagre amounts) during Pol Pot times. As such, their assertions link to important thinking on 'genocide by attrition', which is achieved not only by murder but rather 'several methods, including creating conditions undermining physical and mental health that regularly result in death of part of the group and demoralization and atomization of the remainder' (Fein 1997, pp. 20 and 30). Capital accumulation therefore designates the bodies of evictees as 'surplus populations' of no value except for the land under their feet. The need to understand the co-constitution of direct and structural violence is something emphasized in recent geographical work (Davies, Isakjee, and Dhesi 2017; Loyd 2012; Tyner 2016a) and is also part of the argument put forward in this chapter through its focus on forced eviction (and domestic violence) as embodied terrains of letting live/die. James Tyner (2015, p. 2) writes cogently in this regard that 'practices of dispossession, enclosure, privatization and commodification continue to consign large segments of the world's population to hunger, malnutrition and susceptibility to disease and death. That these practices are not always seen as "violence" highlights the ontological rupture between "killing" and "letting die".'

This ontological rupture and equivalence with genocide is undergirded, second, by forced eviction having a profound influence on the material sustainability of everyday life for ordinary Cambodians (again referred to as citizens). Da and Kravann put forward the notion that everyone was equal under the Khmer Rouge but today forced eviction effects only the poor.[21] As Da explains, forced eviction falls predominantly on vulnerable citizens whose homes and bodies are considered by the government as deleterious to its aspirations to showcase Phnom Penh as a world-class city unoccupied by poverty and the poor.

Third, their narratives offer up contradictory accounts of the strategies that claim to 'secure' and 'develop' a country and its population. In this sense, the narratives of (past) BKL inhabitants question the shibboleths about development and progress by bringing to the fore the injuries they have sustained. Forced evictions are commonly justified as being in the public interest. This is despite continued exhortations of the sanctity of marriage and the negative repercussions of forced eviction for the social reproduction of family in Cambodia (see Chapter 6).

Evidence of the public interest argument for forced eviction can be found on the Skukaku website and in a letter from Keat Chhon, Minister of Economy and Finance and Deputy Prime Minister. He thanked Shukaku for their investment to date.

> Shukaku is a leading Phnom Penh-based real-estate developer committed to helping Cambodia transform to meet the demands of the 21st century. Shukaku was established in September 2001, with the vision to build Cambodia's future by creating modern, sustainable, forward-thinking and socially responsible developments ('About' Shukaku 2016; see http://shukaku-inc.com/about-us/).

> Shukaku is a values-driven organization, guided by our principles in the way we work and the way we think about our developments. As such, we infuse purpose, sustainability and harmony into all that we create ... (Shukaku mission statement 2016; see http://shukaku-inc.com/about-us/).

The Ministry of Economy and Finance strongly hopes – and is confident – you will continue to be a good example and co-operate closely with the government to continue to reduce residents' poverty and contribute to national development (Keat Chhon, cited in Yuthana and Worrell 2013).

With reference to the loss of land and community, Judith Butler and Athena Anthanasiou (2013, p. 2) contend that 'being dispossessed refers to processes and ideologies by which persons are disowned and abjected by normative and normalizing powers that define cultural intelligibility and that regulate the distribution of vulnerability'. In terms of intelligibility, the above examples indicate how Shukaku and the Cambodian government used the foil of national development and claims to 'sustainability and harmony' to justify their actions.[22] As the next section progresses on to show, however, this rhetorical presentation is undermined by the deadly foundations of 'development', which have contributed to the unsustainability and disharmony of homes in BKL and across Cambodia.

Deathly Foundations of 'Development'

In this second section on forced eviction, I sink deeper into the deathly foundations of 'development' manifest in various instances and experiences of suicide, death-in-life, and social death. These multiple forms of death closely align with those I discussed amongst domestic violence victims. Taken together, the data presented on domestic violence and forced eviction justifies and forwards the bio-necropolitical orientation of this chapter and its emphasis on women's experiences of, and domestic interventions in, the 'life-death continuum' (Braidotti 2013, p. 130).

On 22 November 2011, Chea Dara – a 33-year-old wife and mother of two children in BKL – committed suicide by jumping off Chruoy Changva bridge in the city. While holding on to the bridge handrail, it was reported that she called her husband, telling him to look after the children. Her husband recalls, 'She told me that later authorities would force us to move to another place, because they did not agree to give us land ownership – if we didn't have a house, how could we live?' (cited in Chakrya and Doyle 2011, np).[23] Chea Dara's body was later recovered 20 km downstream by the community. In the remembrance film 'Pushed to the Edge' (LICADHO Canada 2011), the video of her public protest and funeral is overlaid with the specially composed song in Khmer 'Have Compassion on Us', which urges authorities 'do not make more fire' and asks that action is taken to ease the injustice of the 'dark and dirty world' that has involuntarily engulfed their lives. A considerable number of BKL residents have become 'ghosted casualities' through 'violent conversion of inhabitant into uninhabitant' (Nixon 2011, p. 152), be this displacement from BKL or displacement from life itself.[24] Suicide is core to the necropolitics of intimate war that can be witnessed at BKL and has been explored though the contemplation of death by domestic violence survivors.[25]

While the remaining community living in the wake of Dara's death asked that the fire of forced eviction not be stoked, the flames were unrelenting. As Sear Naret told me:

> If I did not move myself, the company told me they would submerge my house under the sand, or they would use the bulldozer to demolish it without any reparation. They committed violence to finally force us to leave our house.

Sear Naret describes having a 'huge house' in which she was able to run an intergenerational household prior to her forced eviction. From the house that she had lived in since 1992, Sear Naret, in her fifties, also sold rice and soft drinks to earn money to send her children to school. She suffered from intimidation by 'third hand gangs' (*bat dai ti bai*) who covered their faces with masks, wrapped scarfs (*krama*) around their heads, and carried guns to repeatedly threaten violence if the family did not accept the below-market-value compensation. Sear Naret resisted accepting the compensation but the developers began directing the sand pump towards her house in a form of attrition warfare designed to wear down the family's resolve to reject compensation. She was plagued by insomnia, the pump whirring through the night as she tried to sleep fending off dark thoughts of drowning. Waiting for, and anticipating, the 'fast violence' of eviction, much like for Orm, was experienced as a slow violence (Hyndman 2019) and form of 'stationary displacement' (Davies 2019, p. 8). As Jayson Richardson et al. (2016) identify, for women in Cambodia who have endured forcible eviction, overwhelming sadness, compulsive worry, difficulty in sleeping, inability to concentrate, and emotional outbursts are commonplace. In Sear Naret's case, she eventually lost both her home and business and only received US$8500 in compensation (she estimates that a similar plot of land in the area is estimated to cost upwards of US$80 000). She told me crying:

> Honestly speaking, when they submerged my house, I wanted to sit in my house waiting for them to kill me. I collected all my clothes around me and intended to set fire to myself. If my house was gone, I would agree to die in my house … actually, I wanted the government to see their actions towards their citizens. Yet my mother in her elder age begged me to move out. One side seeing my mother, and other side seeing my house; it was the hardest decision to make and I moved away from my house. Preferably, I would wish to die rather than living through such inhumane treatment. I have no money, my family has broken down, and my husband has a new girlfriend. How can I be expected to live like this? Frankly, if the government killed me, I would not be as angry as this. It would be less suffering for me … for women, if our houses are gone, it is as if we are dead. Our children would have nowhere to live in. As you know, our houses are our lives … we are more hurt and negatively affected than men when we lose our house … women pay more attention to their kitchens, shelters, and family members; as a result, we are more hurt and negatively affected when we lose our houses. I am waiting for the day I die since there is no

happiness and warmth; my family is torn apart! This is the most torturing experience of my life! My life has become meaningless!

Much like for domestic violence victims Nakry and Sopotevy, Sear Naret represents death as a likely release from her inhumane treatment. Her narrative of planned immolation complicates notions of fire in Cambodian society only as slow violences lived by domestic violence survivors. Much like the fire which was used to burn Orm's home into oblivion, Sear Narat's intention to set fire to herself is a faster, yet alarmingly painful and desperate, act of erasure to clear the ground for new growth. While in the former example this new growth takes on a capitalist inflection with the building of tourist infrastructure for profit-making, in the latter case fire offers the restorative possibility of hope and the 'good life' for evictees. In this sense, Sear Narat shares a similar sentiment to an Afghan interviewee in the Calais study by Thom Davies et al. (2017, p. 18) who remarked, 'A quick bullet through the head in Afghanistan would be better than this slow death here.'

Very similar to the experiences of domestic violence survivors told earlier, Sear Naret contextualized for me her 'death-in-life' (Mbembé 2003, p. 21) marred by marital breakdown, financial destitution, and home loss. Staying alive was a 'death-in-life' borne from an altruistic concern for, and onus placed incumbent on her by her mother and children. Her reference to her 'torn family', which has lost 'warmth', also has echoes in other BKL residents' existential concerns of uprootedness and struggles for self-determination in which the 'torn body' is 'hewn in a thousand pieces and never self-same' (Mbembé 2003, p. 27). Sear Naret (much like Orm) also described her miserable life as akin to 'duckweeds' – an aquatic plant common to Cambodia and an evocation of pond life, a contemptible or worthless people. She also believes that the forced eviction has affected her disproportionately as a woman, stating frankly, 'if our houses are gone, it is as if we are dead'. This was a response shared across the women's interviews, including Yorm Bopha's, which I introduced in Chapter 1. As such, forced eviction is registered as an attempted removal of life's meaning that women value. As Sear Naret articulates, her expulsion from home has been accompanied by her husbands' infidelity and the breakdown of their marriage. Their combined loss is a form of gender-based violence that inflicts emotional harm and undermines their confidence and trust in marriage.

For many women, when their ability to fulfil domestic ideals and identities are compromised to such an extreme, a pervading sense of worthlessness can emerge. Forced eviction is an injurious form of capitalist accumulation that inflicts emotional and psychological trauma and wounding on women and their ability to ensure the continuity of the domestic sphere. There is mounting evidence that it is women who disproportionately have to cope with the burdens of forced eviction and the precarity that spirals from it (Brickell 2014a, 2016; Fernández Arrigoitia 2017; Lamb et al. 2017). COHRE (2010, p. 8) argue, for example, that women 'are most often charged with taking care of the children and family before, during and after an eviction, and for providing a sense of stability of home'. As the Committee on the Elimination of Discrimination Against Women (CEDAW)

(2013, p. 10) remarked, the ruling Cambodian Government needs to more fully 'Recognize that forced evictions are not a gender-neutral phenomenon, but that they disproportionately affect women, and take immediate measures to protect women and girls from further evictions.' When homes are forcibly evicted, the onus remains on women to remain the familial anchor. The altruistic politics of the crisis ordinary and the centrality of marriage and its (potential) breakdown are thus of pivotal importance across the domestic violence and forced eviction cases.

The sense of domestic and familial dislocation that women hold primary responsibility for counteracting was particularly acute for BKL residents who moved to resettlement sites, including Damnak Trayoeng located 20 km southwest of the city.[26] Trapaing Anhchanh is another resettlement site that I visited to meet with five families from another eviction site in central Phnom Penh but which speaks to BKL residents' reported experiences.[27]

A verdant green, highly manicured landscape of leisure, Civica Royal Cambodia Phnom Penh Golf Club (Figure 4.5) sits alongside the dusty parched terrain of empty plots and makeshift homes (Figures 4.6 and 4.7) of the Trapaing Anhchanh resettlement site. While the website asserts (in awkwardly translated English) that the 'club surely leads Cambodian's politics and economy circles to a brand-new world of golf recreation' and its friendliness to the environment 'is one of the features of our club's humanity', one only has to take a *tuk-tuk* down the bumpy road alongside to find the antithesis of a habitat for humanity. Housing, health, and education facilities are in an 'appalling state' and families have been moved such a debilitating distance away from the city that their livelihoods and support networks have been destroyed (Zsombor and Phorn 2014). Too often in

Figure 4.5   Civica Royal Phnom Penh Golf Club, 2013. Photo: Katherine Brickell.

Figure 4.6   Vacant plots at Trapaing Anhchanh resettlement site, 2013. Photo: Katherine Brickell.

Figure 4.7   Typical makeshift housing at the Trapaing Anhchanh resettlement site, 2013. Photo: Katherine Brickell.

Cambodia, as elsewhere, 'Resettlement has been so poorly planned, financed, implemented, and administered that these projects end up being "development disasters"' (Oliver-Smith 2009, p. 3).

Such is the level of international concern over BKL and other evictions that in May 2012 the then UN Special Rapporteur to Cambodia underwent his seventh mission specifically focused on ELCs and human rights. Although Surya Subedi was careful to preface that he 'understands that Cambodia, as a developing country, wishes to capitalize on its land and natural resources with a view to promoting development and bringing prosperity,' his report states 'that the human cost of such concessions has been high' (United Nations General Assembly 2012). Resettlement is a poignant exemplar here. Eviction resettlement sites are 'abject spaces' in which 'increasingly distressed, displaced, and dispossessed peoples are condemned to the status of strangers, outsiders, and aliens' (Isin and Rygiel 2007, p. 181). Saroeung told me of her experiences, which speak to this crippling abjection:

> On Sunday 25 November, my husband left us forever. He did not get sick or show any signs that he would go. At around 4 a.m. he woke up and got dressed to go to work as normal, but he realised it was too early, that it was still dark outside. Hence, he went back to sleep and covered a blanket on his body. A few minutes later, I saw saliva all over his mouth. I was so scared and asked my son to call my mother. I did not know that it was the last time I would be with my husband … I just thought my husband was too exhausted after work. In the morning, he'd worked in the water uprooting plants, in the afternoon he'd carried all the packages to the truck. He used up his energy to earn money to pay for the hospital when I was in labour with our baby. He was so worn out. He'd also worked as a security guard for three months, but the golf club did not pay him his due salary. Finally, he changed to work in the construction site until his death.

Saroeung was a 31-year-old widow when I met her cuddling her new baby in Trapaing Anhchanh. A mother of four children, she had lost her husband three months ago. With rising debt and joblessness after their eviction from Russey Keo, her husband Atith had taken on multiple low-paid jobs locally to try and make ends meet (and was in the case of the golf course left unpaid). Atith's unexpected death was attributed to the heavy manual labour and long hours he was committing to pay for the birth of their baby at hospital. The family, like many in the resettlement site, were also living on extremely limited amounts of rice. As Saroeung admits, he was 'worn out'; Atith succumbed to a 'slow death' (Berlant 2007, p. 754) through his working life and his starved body eroded down through the embodied precarity of forced eviction. While paternal irresponsibility was a theme that cross-cut the domestic violence and forced eviction interviews I undertook with women, altruism and sacrifice are not exclusively female behavioural domains. For male breadwinners who contribute to the financial sustainability of their households, the costs of their survival-work in situations of forced eviction can be fatal. Bodily depletion and the erosion of the protective infrastructures of domestic life have gendered repercussions for men as well as women. Indeed, Atith's death shows the need to go beyond what James Tyner and Stian Rice (2016, p. 48) surmise as the 'moral partition' between 'killing' as an action and 'letting die' as a lack of action. The inability to survive transcends the separation of 'direct' and 'structural' violence.

For example, having borrowed money from a local lender to pay for the materials and construction of their house, Saroeung explained her next dispossession was imminent given her inability to repay; debt again was part of the intimate war on familial life that forced eviction extols over time. The deleterious trajectory of debt-taking in Cambodia (Bateman et al. 2019) is particularly acute for evictees (and, as earlier, domestic violence victims too). These concerns about debt across the two case studies mirror those of other scholars who point to the production and reliance upon precarious work and life to achieve the rapid growth of microfinance currently being witnessed in Cambodia (Green and Estes 2019; see also Bylander 2017; Green and Bylander forthcoming). This precarity was felt by Chavy whose interview took the view prevalent in the wider data set that it was men who were largely to blame for household debt:

> I can tell you that, in general, men in Cambodia lack a sense of responsibility to deal with their family's financial situation. Probably, when the husband goes out to work, he meets friends to drink and get drunk, so he does not worry about home. All the time a wife is trying to protect the husband's face by trying to deal with the debt alone because she is aware that if her husband has to deal with the money lender, he may not be able to control himself, and it can lead to violence. This can be one of the reasons that the wife often hides from the husband that she borrows the money from a lender since she does not want the problem getting bigger. If the husband knows that they are in debt, it is more likely that they will have problems with the money lender when he/she comes to collect the money back …. Consequently the problems are getting worse, and women's life at home becomes even more stressful. Women worry so much about the security of their children and husband … about food running out … as a woman who spends her life creating a home, she wants to take care of her family and stay there for the rest of her life without relocation. When we were evicted, it was extremely painful and it really hurt emotionally. That place was the house I used to live in. Women understand the reality, that we do not spend a day to build a house, that takes a very long time. It seems like our future is gone with the eviction! Our land and houses are our lives.

The excerpt from Chavy's interview connects to an earlier part of the chapter on women's experiences of their violent husbands' pursuits outside of the domestic sphere, including drinking. She contrasts men's extra-domestic activities and propensity for violence with women's domestic diligence in trying to deal with familial debt.[28] In Chavy's case this meant taking on extra debt to cover original loans without her husband's knowledge. Chavy's suggestion then is that it is women who carry the burden of anxiety and stress of over-indebtedness as part of the crisis ordinary steeped in spiralling debt. The emotional labour of survival-work includes the altruistic load of sustaining life amidst the attritional after-effects of forced eviction, and includes mounting debts. As Maryann Bylander's (2017) research has established, this is a phenomenon that has only intensified over time and that has seen a growing reliance on loans taken to service existing debts. This reliance is exacerbated by the alcohol-soaked homes that other women across the domestic violence and forced eviction research describe as habitual fuel in the intimate fires they are contending with.

The rapidity of the home's material dismantling through forced eviction contrasts with the slow violence of its perhaps impossible remaking under such conditions of debt and precarity. As Chavy spells out, 'we do not spend a day to build a house, that takes a very long time'. 'To restore any sense of domesticity', John Berger (2007, pp. 122–123) contends, 'takes generations'. In this sense, forced eviction is a slow violence that manifests itself in the truncated capacities of women to maintain or rebuild a stable home core, not only to their self-esteem but also the very survival of their family. Forced eviction, and similarly domestic violence, therefore 'does not simply drop' victims 'into a dark valley, a trying yet relatively brief detour on life's journey. It fundamentally redirects their way, casting them onto a different, and much more difficult, path' (Desmond 2016, p. 30).

This redirected path and slow violence was also evident for evicted BKL resident Phorn Sophea, who moved to a resettlement site after heated and protracted consideration. Buoyed initially by the prospect of gaining formal title to her land, she had attended a meeting at a local pagoda during which assurances were given that BLK residents who had peaceful possession of their land prior to land registration would qualify for official designation under the Land Management and Administration Project (LMAP). Phorn Sophea was promised furthermore that a designated adjudication team would oversee this. The LMAP was set up in the same year by the World Bank and other major donors that aimed to 'reduce poverty, promote social stability, and stimulate economic development' via an efficient and transparent land administration system (see Flower 2016 for a full overview and analysis). While nearly a million titles were issued across the country, those for disputed land were often refused (Strangio 2014), including in BKL.[29]

Under Cambodia's 2001 Land Law (Article 30) '[a]ny person who, for no less than five years prior to the promulgation of this law, enjoyed peaceful and uncontested possession of immovable property that can lawfully be privately possessed, has the right to request a definitive systematic registration certificate' (Royal Government of Cambodia (RGC) 2001). Yet in BKL possessionary claims that residents had right to were undermined by a 'legal conjuring act' to ensure elite territorial control (Strangio 2014, p. 161; see also Flower 2019a and 2019b for a wider analysis on legal tools of spatial exclusion in Cambodia). After the transfer of the lease to Shukaku, in 2008 the government went on to pass a sub-decree that changed BKL from state public property to state private property. This was a key manoeuvre to deny residents title because 'the mere assertion of state ownership of the land was sufficient to halt the determination of private ownership of the land' (Fitzpatrick 2015, p. 76).[30] Law had become a mechanism for 'punative urbanism' (MacLeod and McFarlane 2014) and 'administrative violence' (Spade 2015).[31] Occupation was therefore superseded by the word of law (Springer 2013a).

With Shukaku setting up an office in the community, Phorn Sophea's family was offered US$8500 in compensation. Although located 70 metres from the lake's shore, her home was flooded and she and her family regularly slipped and injured themselves wading through their home. Fearful that she would soon be left without a property to gain compensation for, and scared for her children's safety with electricity cables and water meeting all too often, she accepted compensation. Describing her family as a once middle-class one in BKL, owning publishing machines and running a successful business, Phorn Sophea went on to chart the downward trajectory she has been desperately managing since. The family have been financially unable to purchase land closer to the city; her oldest child has dropped out of school; her two younger children now work as waiters; and her husband has become a *motodop* taxi driver. She was also consumed with regret for leaving the house that she loved and regret for the misery and torture that she encounters today. Phorn Sophea inhabits what Lauren Berlant (2011, p. 7) might describe as 'a moment of extended crisis, with one happening piling on another'. While I opened this part of the chapter by characterising forced eviction as an 'open annihilatory battle', it is also a form of 'slow death' given the 'structurally induced attrition' (Berlant 2011, p. 102), which takes its toll on peoples' lifeworlds and bodies over time.

Phorn Sophea's narrative demonstrates the intimate incursions of necrocapitalism into the entirety of everyday life, rendering vulnerable the security of her family's finances, her marriage, her ability to mother and educate, her self-esteem, and hope. While in her interview she cited the proverb 'plates in basket will rattle' to suggest that for those who live in the same household, collisions and conflicts between one another are to be expected (Brickell 2012c), Phorn Sophea is clear that the magnitude of their marital and parental disputes has been amplified by the impossible decisions she has been forced to make about the long-term future and well-being of her family in the context of forced eviction. Her experiences and those of other women I interviewed demonstrate how the strained or broken marriages of forced eviction become part of its metaphorical rubble. Emphasising this point further, Phorn Sophea told me, bitterly crying, that she was forced by authorities to tear down her own house so there was no chance she could return. Her story not only illuminates the material nullification that eviction in Cambodia can bring about but also the depleted bonds of affection and sense of warmth that women are charged with maintaining under circumstances of intimate war.

Having said this, while I have focused primarily on women's disproportionate burden of forced eviction in this chapter so far, my research in 2014 explored the extreme strains that BKL men also faced. Chankrisna was in his mid-sixties when I met him.[32] He moved to BKL in 1993, bought land (without formal title) and built a house on it with 12 adjoining rooms, which were rented out. Additional income was derived through ad hoc construction labouring. His life in BKL came to a crashing end as his home was bulldozed in September 2011 after it was arbitrarily disqualified from 12.44 hectares of land given to remaining families. Eight homes and businesses were demolished without warning by two bulldozers accompanied by around 100 armed riot police and security guards (Amnesty International 2011). Meanwhile the home of Chankrisna's son close to the mosque in BKL was submerged by flood water.

Homeless and dispossessed from land, which he estimated to be worth US$100 000, Chankrisna bought a plot of land for US$5000 and built a house on it using salvaged wood from his former home. The land was purchased in close proximity to the capital's airport. It soon transpired that its relative affordability could be explained by the risk of eviction the area faced. While lesser known than BKL, in November 2012 the area gained its own notoriety. In advance of then President Barack Obama's arrival to the 21st ASEAN Summit and 7th East Asia Summit, residents painted the letters 'SOS' on their rooftops in a plea for help. Eight residents from the community were arrested.

Chankrisna's interview was inflected with continued uncertainty about another possible eviction. He contrasts the happiness of his 'good life' in BKL with the 'unimaginable' misery of his current situation. A well-known figure in the community who 'was able to participate in ceremonies and enjoy life', the impact of the forced eviction is not limited to the loss of land and property but is an intimate war that negatively impacted on his ability to practice and maintain affective ties to the community. Forced eviction had created 'root shock', a 'traumatic stress reaction to the destruction of all or part of one's emotional ecosystem' (Fullilove 2004, p. 11). Chankrisna's interview painfully articulated forced eviction as 'a form of death-in-life' (Mbembé 2003, p. 21) in myriad ways. He explained:

> I understand that women have suffered much from this eviction. But in the meantime, men have also suffered because men are income earners. I now earn less than zero after I relocated to another area. Sometimes I am called upon to weld iron but I am not paid enough to live on. Here [in BKL] I earned around US$200/month. I cannot describe how I live now [tears streaming]. I was able to participate in ceremonies and enjoy life. My life after relocation is unimaginable! When I see a crab or can on the road, I will collect it to eat. This development has left me in agonizing pain. Honestly speaking, I am thinking of burning myself in front of the Municipality of Phnom Penh. I want to use gasoline to burn myself. I could never imagine that Cambodia would develop in this way! Again and again, I think it would be better I died …. I have nothing to fear any more …. I am not afraid of death.

Chankrisna's interview, and the case of Atith discussed earlier, underscores the concern of many women that forced eviction is not just a loss of shelter but also a loss of livelihood that negatively effects men and women (see also McGinn 2013). It suggests how forced eviction has challenged his ability to live up to idealised forms of hegemonic masculinity given that the earning of money is also about the earning of respect and dignity. Chankrisna describes looking for scraps of food on the roadside and the time and energy consuming nature of this everyday task of securing food and other life necessities. This is a structural violence of poverty, displacement, and hunger, which is 'usually concealed within the hegemony of ordinariness, hidden in the mundane details of everyday life' (Merry 2009, p. 5). Chankrisna's experiences of scavenging reinforce the significance of the embodied terrain of letting live/die that I discussed earlier in relation to Da and Kravann's interviews, which touched on matters of hunger and abjection.

Forced eviction has a physical and psychologically damaging impact that is not limited to women's bodies, especially when it comes to unemployment and the loss of identity in the stage of 'reconstruction' for men (Meertens and Segura-Escobar 1996). This includes the envisioning of death shared with other participants in this chapter. Chankrisna regularly reflects that 'I think it would be better if I died' and references thinking about self-immolation as a means of communicating his despair and pain to authorities. Suicidal behaviour in this context, 'is best seen as a cry of pain – a response elicited by this situation of entrapment – and only secondarily as an attempt to communicate or change people or thinking in the environment' (Williams 1997, p. 139). Death, Mbembé (2003, p. 39) argues, can be agentic as 'a release from terror and bondage'. In Cambodian culture, however, there is an important distinction made between 'good' and 'bad' deaths: the latter being people who commit suicide, are murdered, have fatal accidents, or die in childbirth (Ledgerwood 2012). In Cambodian villages, for example, the ghost of someone who was murdered or committed suicide is particularly feared (Harris 2005, p. 59). The fact that this was not mentioned in any of the interviews I conducted might indicate the depth of despair felt; that nothing could be worse and thus suicide is a worthwhile risk. Writing on the meaning of suicide in Buddhism, Karma Lekshe Tsomo (2006, p. 134) explains: 'To commit suicide as a means of alleviating mental or physical suffering is misguided, because the suffering in the next existence may be even greater, especially as a result of committing suicide.'[33] Nevertheless, both domestic violence and forced eviction narratives reveal how death is a performative prospect or act of dying to prove life.

A sense of entrapment and hopelessness ran through Chankrisna's interview and nowhere more so than when he began talking about his wristwatch (Figure 4.8):

Figure 4.8   Chankrisna's watch, 2014. Photo: Katherine Brickell.

I wore this wristwatch while I was protesting at Freedom Park, and I was asked why?[34] I replied that as soon as politics changes, I will stop wearing it. I worked in the village, and then I was forced to leave my position. I am reminded that because of this wristwatch, I have lost my US$100 000!

When Chankrisna lived in BKL he was an official for the ruling Cambodian Peoples' Party (CPP) and was gifted the watch (see Hughes 2006 for a fascinating discussion of the symbolic and coercive dimensions of gift-giving to individuals by the CPP). The watch displays a headshot trio of Prime Minister Hun Sen in the middle flanked by now deceased CPP President Chea Sim and President of the National Assembly of Cambodia, Heng Samrin. His interview squarely apportioned blame to these ruling elites for the forced eviction of BKL. For this reason, he goads himself by continuing to wear the CPP watch, a form of self-harm that he rationalized as a way of communicating the betrayal and injustice he feels. Forced eviction is felt, embodied, and resisted through gendered bodies: it is a 'loss of confidence about how to live on, even at the microlevel of bodily comportment' (Berlant 2011, p. 16). The guttural sense of wrongdoing haunting Chankrisna compels him to wear the watch almost like a barb. Crisis surrounds and envelopes life (Anderson 2016) and his wearing of the watch commands that his eviction and its long-term impacts be viewed as an ongoing, rather than finite, harm. Forced eviction is a not a singular event in Chankrisna's life that necessarily needs a reminder around his wrist. While the precarity of the first eviction continues to unfurl, the prospect of a second life-churning dispossession compounds the seemingly inescapable 'severe death immersion' (Lifton 1991, p. 503) that his life has become, and the watch embodies. It is also a timely barometer of how the displacing tendencies of elites are tightly wrapping around and constraining the homes, bodies, and futures of individuals, families, and communities trying to survive in crisis ordinary Cambodia.

Conclusion

In this chapter I have explored domestic violence and forced eviction as intimate wars of attrition that Cambodian women are experiencing and navigating. While they may be sidelined as more diffuse and less spectacular forms of violence than classical war, I have positioned them at the frontline of debate and women's everyday survival-work. Analysing domestic violence and forced eviction in tandem reveals the weight of survival-work that their victims face, to the extent that many women do not foresee living free from violence without the release of death. Rather than of short duration and minimal scale, crisis for the women I interviewed across the two cases is a 'fact of life' that is lived 'as a fact in ordinary time' (Berlant 2007, p. 760). In feminist writing on disaster-induced displacement in the Philippines, Maria Tanyag (2018, p. 633) observes that 'there may be a widening of pre-existing gaps between the intensified provision of care and the contributions to sustain the very bodies that meet heightened care demands precisely because this is when gendered expectations of altruism and self-sacrifice operate the most'. The crisis ordinary prolongs these intensified responsibilities and suspends both the hope and capacity of women to address the gendered inequities of social reproductive work they do to sustain home life in such precarious circumstances.

Together domestic violence and forced eviction also bring into sharp relief the bio-necropolitical production of disavowed subjectivities, marriages, families, and intimacies pursued through the destruction and/or denial of home. I have established, for example, how home is being instrumentalised in the attrition war against Cambodian women to conform and submit to patriarchal power. The loss of home is of paramount concern given the significance of the domestic sphere to women's identity and power in Cambodian society. Indeed, the 'multiple practices of dying' (Braidotti 2011, p. 335) discussed in this chapter bear record to corporeal violences against their bodies and the homes they are tasked with (re) making. They are bodies and homes 'sharing the common features and fates which affect the sense of self' (McDowell 2007, p. 93). In so being, they collectively inhabit the 'zone of ordinariness' (Berlant 2011, p. 96) in which the interrelationships between gender, multitudes of violence, and the crisis ordinary become perhaps best known. In this context, the chapter has also demonstrated how (re)marriage and domestic materiality offer important means for undertaking associated survival-work. They are deemed vital to women striving to mend personal and community projections of morality, dignity, and identity; my point is that any focus on domestic violence and forced eviction does not, and should not, preclude an equivalent lens on home-making.

The chapter has also reinforced my argument that domestic violence and forced eviction should be understood as forms of gender-based violence. While in the first half of the chapter I showed the veracity of physical and sexual violence located within marital relationships in the form of domestic violence, in the second half I provided a brief mention of feticide as a direct and deadly violence perpetrated by security forces against forced eviction activists (see also Chapter 6). Psychological violence in the form of intimidation, threats, and isolation were also experienced across the research on domestic violence and forced eviction. The emotional harm and pain experienced from the material and/or metaphorical breaking apart of home took its toll further. This was enacted through physical violence against the home itself, be this its disassembly and removal by an estranged (ex-)husband or the force of a sledgehammer reluctantly exerted by an occupant forced to destroy her own home in BKL.

Economic precarities of home life in crisis, including unsustainable debt, compound these losses for women who have experienced either domestic violence and/or forced eviction. Domestic violence victims are shown to endure economic violence in the form of denied control and access to financial resources. This includes prohibition to work, financial deception and indebtedness, withholding of money and child maintenance payments, and the destruction of jointly owned assets (such as the home). Meanwhile, the narratives of forced eviction victims show how they see themselves as the 'collateral damage' of necrocapitalism in which their lives and livelihoods are ghosted to service the economic logics of contemporary Cambodia that dispossess them and then enrols them in greater debt. Both domestic violence and forced eviction are therefore prevalent gender-based instantiations of economic violence in contemporary Cambodia. They evoke and cement feminist assertions that with capitalism comes an intensification of violence against women (Federici 2014; Mies 2014).

These intensified yet ordinary violences have diverse temporalities and horizons ranging from slow attritional violence to a fast annihilatory sort. However, just as life and death are graduated domains in Cambodia, so too are these velocities of violence mutually imbricated. Interlaced through the chapter, I have explored how women's lives are inflected with a gruelling longevity of domestic violence and forced eviction with a close equivalence to an intimate war of attrition. Attrition warfare in military circumstances is characterized by belligerence and the infliction and sustaining of injury and destruction against the enemy across time (Latham 2002). Financial control, sleep deprivation, shame, and stigma, are all tools variously weaponised across time to ensure women's acquiescence to the normalcy of violence in everyday life. Their endemic potency is punctuated by episodic direct violences from a metal torch crashing down on a woman's head to a bulldozer wrecking a home. Trauma thus results from both fast and slow violences (Pain 2019). The haunted and starved lives of families moved to resettlement sites in the case of forced eviction, for example, elongates such violences in space/time and are indirect methods of intimate warfare waged against evictees' bodies, including those of men. Both domestic violence and forced eviction demonstrate that as a spatially and temporally imbued concept, slow violence compels scholars 'to include the gradual deaths, destructions, and layered deposits of uneven social brutalities within the geographic here-and-now' (Davies 2019, p. 2).

Yet as the chapter has clearly shown, analysis of the 'geographic here-and-now' must also include historical junctures and violences that are lived through the still present. The 40-year arc since the reign of the Khmer Rouge has seen the (continued) unfolding of these intimate wars, which add to the precarity of current-day householding. What revealed itself through the chapter is the asymmetry of explicit reference to the previous Khmer Rouge period across the domestic violence and forced eviction case studies today. While BKL women and advocacy groups publicly and strategically mobilise the country's past genocidal history for its analogous potential in the here and now, the interviews I undertook with domestic violence victims made far fewer allusions and connections to the trauma of the period and its legacies, a subject on which many women remain silent.[35] Apart from being a response to its intimate trauma, silence is used by Khmer women, Majorie Muecke (1995, p. 44) writes, 'as an implicit cultural strategy for survival in their communities ... they serve the Cambodian cultural mandate to appear virtuous'. As such, domestic violence and forced eviction are protracted violences and crisis temporalities, which have durable yet differentially voiced traces in contemporary Cambodia.

In both cases, Cambodian women are charged with being 'resilient subjects' who take on 'the responsibility to adapt to, or bounce back from, inevitable shocks in an unstable world' (Anderson 2015, p. 61). The 'unstable world' and the place of home within this ontology of rupture is one that meets the criteria of the crisis ordinary as they manage 'fire in the house' with its multiple origins and fuels that undergird the growing normalcy of uncertainty. In situations of domestic violence, for example, the chapter has illustrated how women are kept in place by expectations and pressures of altruism to elevate concerns for their children and spouses' well-being above their immediately safety. In rather ironic contrast, women in BKL have been dispossessed by forced eviction despite impassioned vocalizing of their needs and desires to uphold their normative responsibilities for mothering (see Chapter 5). Taking women's experiences and viewpoints seriously means framing these apparent acts of conformism as intelligible ways to cope with constrained agency.

Finally, in this chapter the twin study of domestic violence and forced eviction has shown how a bio-necropolitical reading of contemporary Cambodia can bring to the fore 'how contemporary forms of rule enact a subtle form of death-politics: the production of necropolitical geographies that render some places deathworlds whose slow forms of violence do not elicit shock or emergency response' (Anderson et al. 2019, p. 7). While in this chapter I have focused on everyday, mostly private, acts of resilience that sustain women's survival but with limited political impact, the next chapter turns to public efforts to bring about change in women's lives affected by the intimate wars of domestic violence and forced eviction set out here.

Endnotes

1 It is common for older women in Cambodia to spend time in temples accumulating merit (*bon*) for the next incarnation.

2 Unregistered customary marriage is the norm, particularly in rural Cambodia where many still live under informal systems of law. The formal legal system has only made 'cursory incursions' into daily life (United Nations Development Programme (UNDP) 2007, p. 1). See Katherine Brickell and Maria Platt (2015) for further information on distinctions between types of marriage in Cambodia.

3 In rural Cambodia especially it is common amongst older participants to use a thumb-print signature, given low literacy levels.

4 While ACLEDA started out in 1993 when it began providing microcredits to war victims, it is now Cambodia's major commercial bank.

5 Microfinance began in Cambodia under a 'humanitarian NGO-driven model' (Bateman 2017, p. 3) in the early 1990s, designed to provide opportunities for demobilized soldiers to start up small-scale enterprises as a means of social reintegration following decades of war. However, by 2000, the state-run regional microlending projects had been centralised into a single profit-oriented entity, officially becoming a private commercial bank in 2009. The shift to a profit-driven model of microfinance necessarily entailed interest, thus loans were structured to increase surplus extraction from the indebted. This period saw remarkable growth in the sector, with the number of clients served by Micro-Credit Institutions (MCIs) rising from 300 000 in 2005 to almost 1.6 million in 2013 (Bateman 2017). Furthermore, by 2011, the average lending amount had risen to 139% of the country's Gross National Income (Bylander 2015), indicating unstable levels of indebtedness.

6 According to the Ministry of Women's Affairs (MOWA) et al. (2014), 7% of ever partnered women who reported physical or sexual intimate partner violence (*n* = 3043) reported having been raped (*n* = 216). More broadly, sexual violence is a serious issue in current-day Cambodia and starts from an early age. A MOWA et al. (2014) survey found that 4.4% of females and 5.6% of males aged 18 to 24 experienced at least one incidence of sexual abuse before the age of 18 – with neighbours found to be the most common perpetrator of the first instance of sexual abuse. A Partners for Prevention (2013) study also found that out of 8.3% of men who raped a non-partner, 5.2% were multiple perpetrators or perpetrators of gang rape (this is higher than the 2% typically seen in most other sites in Asia surveyed – Bangladesh, China, Indonesia, Papua New Guinea, and Sri Lanka). While gang rape was the least common form of rape in all other country case studies, the exception was Cambodia, where it is more common than non-partner rape by a perpetrator acting alone. Such figures bolster the findings from previous work, which emphasises gang rape (*bauk*) of females as a significant issue (Gender and Development Cambodia 2003). Amnesty International (2010, p. 51) asserts in connection that 'many men's sense of unbridled entitlement to sex is widely considered acceptable, and widespread impunity is reinforcing this sense of entitlement'.

7 Interestingly, research with Cambodian women experiencing domestic violence in the United States also found similar attitudes about obeying and respecting one's husband and maintaining harmony in social relations (Bhuyan et al. 2005). Mainstream domestic violence services in the United States, the study argues, 'often focus their attention on

women's safety: getting survivors away from their abuser. In contrast, women in the current study said that they wanted help resolving family problems with the goal of keeping the family together' (p. 915).

8 See Katherine Brickell (2008) on the relationship between domestic violence and alcohol consumption in Cambodia.

9 Serge Thion (1993, p. 226) explains that in the past 'when dissatisfied with local conditions, Khmer farmers, who do not bury their dead in the ground, could easily dismantle their house, load it on an ox cart and move elsewhere'.

10 This trend goes against Article 74 of Cambodian marriage and family law (Royal Government of Cambodia (RGC) 1989), which places an obligation on both parents to love, bring up, and take care of their children's education, even in the case of divorce proceedings.

11 *De jure* female-headed households refer to women who reside independently of men as a result of non-marriage, separation, divorce, or widowhood, and whose receipt of male support, such as child maintenance, is much less likely (Chant 1997).

12 According to Melanie Walsh (2012) the law regarding physical custody of children upon divorce is weak and in need of revision. While in the majority of other countries resolution involves decision-making by a third party (a court for example) in Cambodia emphasis is placed on parents reaching an agreement. To compound this, there is 'no reference in Cambodian law to situations in which parents reach an agreement' (p. 171).

13 Under General Provisions of 'Things' in the Cambodian Civil Code (2007), the house is classed, however, as 'immovable' given 'An immovable comprises land or anything immovably fixed to land, such as a building or structure' (Article 120).

14 For television news coverage watch the Channel News Asia report http://www.youtube.com/watch?v=n4ErrsOeZCY.

15 See, for example, http://www.youtube.com/watch?v=2-G_wH2wLig.

16 Shukaku is a private development company associated with Cambodian Senator Lao Meng Khin. More general information on urban planning and real estate in Phnom Penh is available (Fauveaud 2017; Percival 2017).

17 While I am cognisant of the critique made by James Tyner et al. (2014, p. 1889), that Democractic Kampuchea (the Khmer Rouge-controlled state) 'did not "neglect" or "abandon" Phnom Penh, but rather reconfigured the city to meet its political and economic needs', the analytical attention of scholarship remains on its initial depopulation.

18 The Campaign was established by the Housing Rights Task Force, a coalition of NGOs working for housing rights in Phnom Penh. The blog is an independent effort to chronicle their efforts. For further information see https://saveboeungkak.wordpress.com/about-the-save-boeung-kok-campaign.

19 See Trudy Jacobsen (2008) for evidence of this argument in relation to gender equality during the regime.

20 Andrew Basso (2016, p. 6) defines a 'displacement atrocity' as 'a type of killing process employed against a targeted population defined by the perpetrators which uniquely fuses forced population displacement and primarily indirect deaths resulting from the displacement and systemic deprivations of vital human needs to destroy a group in whole or in part'.

21 While there was an 'equality of misery' (Yathay 1987, p. xix) that pervaded the pursuit of agrarian socialism, the comparisons made by interviewees do not concede to the clear demarcations of experience that existed during the Khmer Rouge regime. Those who had not lived under the Khmer Rouge-controlled territory prior to 17 April 1975 were considered 'new people', while 'base people', who enjoyed more privileges, were those who had lived in the Khmer Rouge-controlled territory (Chandler 1991, p. 265). Geographical variations also existed; conditions in the north and north-west of the country, where more than a million new people were resettled in 1975–1976, are argued to have been the harshest (Chandler 1991; Vickery 1984). In this way, the Khmer Rouge took 'as one of its pillars that geography was indeed destiny, that where one lived determined one's enemy status' (McIntyre 1996, p. 758).

22 In June 2016, the daughter of Senator and CPP member Lao Meng Khin 'admitted that the relocation of the lake's residents could have been handled better and that flooding issues still persist – though she prefers to talk about the future because Shukaku has completely overhauled their approach towards the development of the former Boeung Kak Lake, which Shukaku now calls Phnom Penh City Centre (PPCC)' (Post Staff 2016).

23 Health researchers who undertook six focus group discussions with students from secondary schools close to Phnom Penh found that suicidal feelings were shaped heavily by the gaze of family and community (Jegannathan, Dahlblom, and Kullgren 2014). In their study, female participants noted the *Chbab Srei* and feeling caught between traditional and modern values as a stress factor. It has been reported that suicidal tendency is a significant issue in Cambodian society. There were an estimated 42.35 suicides per 100 000 in 2011 compared to worldwide numbers, which averaged 16 suicides per 100 000 population in the same year (Schunert et al. 2012). The same report outlines findings from a randomised sample of 2690 adults from nine provinces across Cambodia who were interviewed about mental health and mental disorders, addictive and violent behaviours, suicidal tendency, and traumatic events. Notably, it found an especially high percentage of mental disorders among the female population. Lara Schunert et al. (2012) reported that of suicidal attempts made, 15 were male (1.7% of male respondents, $n = 871$) and 100 were female (5.5% of female respondents, $n = 1809$).

24 Suicide has been identified as an issue associated with home eviction in the United States (Fowler et al. 2015).

25 These intimate portraits sit alongside survey data collected by The Cambodian Center for Human Rights (CCHR) (2016), which shows that 92.8% of women ($n = 600$) reported their mental health had been affected as a result of land conflicts; 46.2% of women had considered suicide; 18.1% had attempted it; and 25% of women still had suicidal thoughts.

26 In BKL alone, out of 4000 original families, January 2014 community records I was shown indicated that only 794 residents remained.

27 Further information on Damnak Trayoeng is available (Mgbako et al. 2010; Talocci and Boano 2015).

28 While Saroeung shouldered the brunt of household survival alone this had not arisen through paternal irresponsibility. Yet the normative discourse of male unreliability continued to be immortalised in the majority of women's interviews I undertook (including Chavy's).

29 In connection to two economic land concessions (ELCs), which took place in 'clear conflict with the laws of the land' (a forestry concession and industrial sugar plantation), it has been concluded that 'While legislation and the institutional architecture in Cambodia seem to offer a reasonable degree of protection to ordinary people (including indigenous groups) and adhere to notions of sustainability, reality on the ground is often dictated by naked power and patronage' (McLinden Nuijen, Prachvuthy, and Van Westen 2014, p. 168). In rural areas of Cambodia displacement is driven by natural resource extraction and ELCs. In urban areas private investment, speculation, real estate development, and urban beautification prevail (Connell and Grimsditch 2017).

30 The Land Law (RGC 2001) was breached in three respects as a result of this switch: 'First, state public property may not be the subject of a lease. The lease was issued before the land was converted to state private property. Second, state public property may only be converted to state private property once its public use is exhausted. The public utility of the lake shore remains as a result of its continued recreational use. Third, the maximum term of a lease over private property is 15 years (not 99 years). At its face, therefore, the Boeung Kak Lake case is simply an example of unlawful land grabbing by the state. However, the core legal design problem was the fact that the mere assertion of state ownership of the land was sufficient to halt the determination of private ownership of the land' (Fitzpatrick 2015, p. 76).

31 For further information on why the BKL eviction violated various areas of Cambodian law, including the 2001 Land Law (RGC 2001) and the constitution (RGC 1993), see The Inspection Panel report (World Bank 2009) and Mgbako et al. (2010).

32 It is notable that Chankrisna choose to use his real name, stating at the start of the interview: 'You can use my real name because I am not afraid of death.'

33 Alberto Vanolo (2016, p. 194) asks geographers working on the spatialities of death to work more on 'traditional religions and these new spiritualities [that] theorize that, despite the collapse of the physical body, other experiences will take form after death'. This follows from the point made by Olivia Stevenson et al. (2016, p. 153) that, 'Whilst deathscapes have been framed within geographical work as incorporating material, embodied and virtual spaces, to date Anglo-American and European studies have tended to focus on the literal and representational spaces of the end of life, sites of bodily remains and memorialization.' As Courtney Work (2017, pp. 390 and 394) writes of Cambodia, 'Boundaries between the living and the dead are created through Buddhist ritual and crossed by the spirits and their living families …. The dead all become spirits, active in the social lives of their communities.'

34 Freedom Park is located in central Phnom Penh. It is a public square that since 2010 was designated as a place for free expression and became a hotbed of opposition protests in the aftermath of the disputed 2013 election.

35 See Chapter 3, however, for formal forums in which women have spoken out about sexual violence encountered during the Khmer Rouge regime.

Chapter Five
(Un)Invited and (Un)Eventful Spaces of Resistance and Citizenship

Introduction

In the previous chapters I described the landscape of crisis ordinary Cambodia and explored women's survival-work through, and moving on from, domestic violence and forced eviction. In this chapter I examine the organised mobilisation of women's rights and rights-talk in respect to each home SOS: first via the formal rights ratification and implementation of domestic violence law (DV law) and second through women's grassroots activism against home dispossession. While these could be distinguished from each other as invited versus invented spaces of intervention (respectively), both bring to the fore complexities of resistance and citizenship that are discernible at the domestic nexus of gender, violence, and survival in Cambodia. Faranak Miraftab (2009, p. 38) defines 'invited' spaces as 'those grassroots actions and their allied non-governmental organizations that are legitimized by donors and government' while 'invented' spaces are 'those collective actions by the poor that directly confront the authorities and challenge the status quo'. Just as she argues that the two types of space – invited and invented – are not binary relations but mutually and dynamically constituted, the chapter provides a nuanced analysis of how SOS responses to address domestic violence and forced eviction are (un)invited and (un)eventful under different circumstances, with particular strategic intents that belie any strict dualistic reading.

Home SOS: Gender, Violence, and Survival in Crisis Ordinary Cambodia, First Edition. Katherine Brickell.
© 2020 Royal Geographical Society (with the Institute of British Geographers). Published 2020 by John Wiley & Sons Ltd.

To interrogate this in more detail, I begin the first half of the chapter by centring my analysis on domestic violence and law as a commonplace language and locale for resistance (Merry 1995). Despite feminist concerns voiced in the late 1980s about the 'malevolence' of law and its embedded 'masculine culture' (Smart 1989, p. 2), there has been an unprecedented growth in the number and scope of laws that extend states' duty to address violence against women, particularly in the developing world. The past century, according to UNWOMEN (2011, p. 118), has 'seen a "transformation" in women's legal rights, with countries in every region expanding the scope of women's legal entitlements'. As of April 2011, 125 countries had outlawed domestic violence (UNWOMEN 2011). Countries with specific domestic violence legislation also increased from 4 to 76 between 1993 and 2013 (Ellsberg et al. 2015).

Honing in on the Cambodian state's parallel turn to legislation, I examine the rhetorical journey and practical mobilization of its first-ever DV law. Cambodia's DV law is, I argue, a form of 'reworking' (to use Katz's 2004 terminology) that forms part, rather than is unsettling of, hegemonic social relations. I contend furthermore that this reworking, in which legal rights are altered but the polarisation of power relations is not (Sparke 2008), means that DV law has limited capacity to achieve emancipatory objectives (resistance) despite the promise invested initially in it by the NGO and international community. As Dean Spade (2012, p. 57) is clear in his work on administrative violence, the assumption that law will change lives for the better 'not only relies on an overly centralized model of power but also misses how law is often one tactic that rearranges just enough to maintain the current arrangements'. The interviews presented are telling of the legal, moral, cultural, and economic onuses that are placed on domestic violence victims to cope and stick with the adverse domestic circumstances outlined in Chapter 4. For this reason, I am conscious to question conventional notions of activism. In the case of domestic violence, I position women's help-seeking, via the law or otherwise, as modest practices of activism given the physical, economic, and emotional dangers this entails in a country with a dysfunctional and corrupt legal system.

In the second half of the chapter I turn to the invented activism of Boeung Kak Lake (BKL) women against forced eviction between 2008 and 2012. By contesting state and World Bank complicity in their evictions, my effective focus lies on a group of women '"out of politics" in any normative sense', but who are 'able to act in ways that allow them to (temporar[ily]) constitute themselves as political subjects' (Tyler and Marciniak 2013, p. 7). In the analysis, I show how BKL women are engaged in insurgent practices, distinct from those of domestic violence victims and their advocates, which have led to their national notoriety. As Judith Butler et al. (2016, p. 3) have written, 'definitions of vulnerability as passive (in need of protection) and agency as active (based on disavowal of the human creature as "affected") requires a thorough going critique' as does the 'epistemic grid laid into that lens', which posits the former to women and the latter to men.

The material transgresses this epistemic grid by concentrating on women's unsanctioned direct action and its extolling of an 'oppositional consciousness' and 'consciousness-building' (Katz 2004, p. 251) more aligned with conceptualizations of 'resistance' rather than the 'reworking' of DV law.

Spanning and connecting the domestic violence and forced eviction case studies, the chapter examines refusals for 'fire' to be harmfully contained to the home and explores the survival-work involved in labouring the crisis ordinary. Both DV law and BKL activism render these violences public through the 'punctual spaces' (Anderson et al. 2019, p. 5) of parliament, the courts, and the street. These spaces are part of the extra-domestic realm that have been 'eventfully' co-opted into women's rights-claiming activities to attract political attention and response. As Ben Anderson et al. (2019, p. 13) explain, the 'aim is to make a situation – to transform an episode, to use Berlant's phrase, or a quasievent, to use Povinelli's – into something that attracts and holds the attention of a public and/or governing authority'. These attempts work to create an 'interval' of action, 'a reversal of how some governing apparatuses function' (p. 5).

While calls for DV law, and the taking part in forced eviction activism, makes imperative 'livable lives, the demand to live a life prior to death, simply put' (Butler 2016, p. 217), the chapter makes it possible to track and ascertain how governmental apparatuses have responded in Cambodia. It shows, ultimately, that the Cambodian government is fundamentally unwilling to sanction women's contestation of domestic violence or forced eviction vis-à-vis the overturning of hegemonic power relations invested with their interests. By this logic, it is favourable for domestic violence and forced eviction to be kept as uneventful as possible, and that any eventfulness on the public stage to be muted, slowed, and, ideally, stopped. These various tactics are used to de-escalate potential crisis for the powerful. In doing so, they stifle women's freedom from gender-based violence as an enduring phenomenon in Cambodia deeply rooted in gender inequality.

Given all of these factors, the chapter continues to forward the significance of the crisis ordinary for training the focus of *Home SOS* on how women manage heightened forms of threat in the context of living. The chapter looks, as a result, to the un(invited) and (un)eventful spaces of resistance and citizenship that have emerged from the extra-domestic realm as an expansive space of intense political and legal significance.

Legislating Against Domestic Violence

In this half of the chapter I combine the life stories of victims Ny, Kesor, and Tola with the wide-ranging perspectives of stakeholders who have professional engagement with DV law and/or DV victims. Drawing on these multiple voices, I divide my analysis into two parts, the first on the long passage of DV law into existence and the second on why and how it has been (im)mobilised on the ground. In contrast to the fast and anticipatory temporalities of state action against forced eviction protesters detailed next in Chapter 6, I show how the temporalities of law-making have been slow, protracted, and, at several junctures, deliberately obstructed.

Harmonising Discord: Cambodia's 2005 Domestic Violence Law

In 2000, multiple NGOs focused on the advancement of women's and children's rights were formally registered (this included GADC and CAMBOW, a coalition of over 30 gender-focused NGOs). The impetus for the country's 2005 DV law (Royal Government of Cambodia (RGC) 2005) initially came about through the collaborative endeavours of these Cambodian NGOs (as well as LICADHO and the GADNET network) in pressurising the Ministry of Women's and Veteran's Affairs (as it was known at the time) to draft a law and include their suggestions.[1] Substantial financial and technical support was also provided through donors including UNWOMEN, UNDP, and the German agency GTZ (now GIZ). According to Rémy Bonneau (2004), working for GTZ, after a first 1996 draft was left idle, a new draft was written in June 2001 but received opposition from NGOs as no penalties were included. The draft was then modified in cooperation with NGOs and penalties were added. A year later in June 2002, the Council of Ministers gave approval for this draft law and, on 2 July, Prime Minister Hun Sen submitted the draft law to the National Assembly. In August 2002 NGOs warned that a raft of flaws remained in the latest draft. Mu Sochua, a forefront campaigner for the law, was concerned, however, that it be passed without further delay or risk 'backfiring' completely given the upcoming general election in July 2003 and expiry of the National Assembly's mandate (Nara and Tithiarun 2002). Arguing that corrections could be made after it was passed, she used the analogy that:

> When we go shopping at the market, we want to buy everything, fish, vegetable, fruits ... [but] we have only one basket and have to put everything in it. I am afraid the basket will leak (Mu Sochua cited in Nara and Tithiarun 2002, np).

The deliberations, hiatuses, and backtracks that came to characterise the passage of Cambodia's DV law are illustrative of competing interests that Mu Sochua was acutely aware of. Similar to women's movements in China, India, and Indonesia, '[l]egal change is a complex and iterative process, in which both state and non-state actors negotiate and bargain over the content of law in the "policy space", bringing different interests and needs to bear' (Cagna and Rao 2016, p. 277). After heated bargaining, on 26 May 2003 the National Assembly voted for modifications to the draft law and adopted only one chapter. More than 200 domestic violence victims under the CAMBOW coalition had planned to gather outside the National Assembly to urge the passage of the law. The Phnom Penh Governor's Office had banned this action, with a LICADHO representative responding, 'we regret that the authorities prohibited our right to express ourselves' (cited in Sokheng 2003, np). At a local NGO office, domestic violence victim Ho Lida sat on a wooden chair with her mouth bound by scarves to symbolize the government's silencing of her and other victims (Naren 2003). Here the closed spaces of decision-making in Cambodia held little prospect of broadening the boundaries of citizen inclusion or even their physical presence outside the National Assembly. Women were uninvited; their collective assembly, like that of their forced eviction counterparts, unwelcomed in the public domain.

Two years later, in June 2005, the National Assembly committee in charge of reviewing the draft law on domestic violence received a new version from the Council of Ministers in which articles already covered by criminal law were removed, including specific punishments for offenders (Rith 2005). With this deletion, Cambodia's DV law was quickly passed by the Cambodian National Assembly and Senate on 16 September 2005, nearly a decade on from its initial drafting (see Box 5.1 for key articles of the law).

As Box 5.1 sets out, while DV law does not contain any criminal penalties, it does state that where the violence amounts to a felony or severe misdemeanour, it can be prosecuted under the Criminal Code (RGC 2007b). Despite DV law not indicating which article(s) of the Criminal Code should be used, the latter law includes Article 222, which states that violence committed by a spouse or partner shall be punished with imprisonment from two to five years. According to the Cambodian NGO Committee on CEDAW (NGO-CEDAW 2016b), there remains however a common assumption that domestic violence is not a crime unless the victim is permanently injured or killed. Further to this, they found evidence that some judges and prosecutors in domestic violence cases say they can only apply Article 222, claiming that other articles which call for longer sentences do not apply because no specific mention is made of spouses and partners. These initial issues are part of a wider set of practical concerns that are elaborated upon in Chapter 6.

Cambodia's DV law was compromised then in comparison to international standards, lacking the comprehensive criminalization approach that NGOs had advocated. The attritional wearing down of penalty provisions took place over the long-time horizon of its passing. Indeed, the strategically harnessed temporal dimensions of law-making and their gendered politics offers an important avenue to think about DV law in Cambodia. Gratitude for its very existence has been simultaneously belied by the undercutting of its potency in practice. Article 4d of the Declaration on the Elimination of Violence against Women (United Nations General Assembly 1993, np) recommends, for example, that Member States '[d]evelop penal, civil, labour and administrative sanctions *in domestic legislation* to punish and redress the wrongs caused to victims' (my emphasis).[2] By eschewing such recommendations, DV law had been 'amputated' according to the female

Box 5.1   Key articles in The Law on the Prevention of Domestic Violence and the Protection of Victims (RGC 2005)

Scope

Article 2: Domestic violence is referred to [as] violence that happens and could happen towards: (1) Husband or wife (2) Dependent children and (3) Persons living under the roof of the house and who are dependants of the households.

Article 6: Tortures or cruel Acts include:

- Harassment causing mental/psychological, emotional, intellectual harms to physical persons within the households.
- Mental/psychological and physical harms exceeding morality and the boundaries of the law.

Prevention and Protection of Victims

Article 17: To participate in the implementation of the penal procedures in effect, the authorities in charge cannot intervene to reconcile or mediate criminal offences that are characterized as felonies or severe misdemeanors.

Article 19: Any domestic violence [in] which the criminal offences are characterised as felonies or severe misdemeanor shall be subject to a criminal suit, despite the violence [being] already over. The criminal complaint shall be made in the form as stated in this law and in accordance with the law on penal procedures in effect.

Article 26: For the offences that are the mental/psychological or economic affected violent acts and minor misdemeanors, or petty crimes, reconciliation or mediation can be conducted with the agreement from both parties. The household members can choose any way by requesting parents, relatives, Buddhist monks, elders, village chiefs, and commune councillors to act as the arbitrators to solve the problems in order to preserve the harmony within the household in line with the nation's good custom and tradition in accordance with Article 45 of the Constitution of the Kingdom of Cambodia.

Article 27: The courts shall try to reconcile the violence disputed parties under the condition that it is in response to the wishes of the household members. While reconciling and mediating, the courts shall avoid putting pressures on the party who refuses to go along with each other or forcing any party to reconcile, or forcing to come into an agreement without the agreement from the two parties.

Penalties

Article 35: Any acts of domestic violence that are considered as criminal offences shall be punished under the penal law in effect.

Article 36: Criminal prosecution shall not be possible if there is a request from a victim who is an adult due to the offences [being] minor misdemeanors or petty crimes. In case domestic violence has been repeated again in violation of the penal law, the courts shall charge the perpetrators in accordance with the penal procedures, even if there is a request from the victims again.

director of a Cambodian human rights NGO I interviewed in 2014. Just as women's homes and bodies are deemed truncated – quite literally – when they fail to control 'fire in the house', the removal of penalty provisions in the country's DV law had trodden its own severed path. From an international human rights perspective too, the DV law contravened Article 4 of CEDAW, which prescribes that 'States should condemn violence against women and should not invoke any custom, tradition or religious consideration' (United Nations General Assembly 1993, np). The extra-domestic home had become embroiled in an intimate geopolitics of law-making, which brought into conflict the respective standards and rationalities of the global human rights regime, NGOs, and those of many Cambodian government lawmakers.

The legitimacy given to DV law as a penalty-free dictate, in which logics of inalienable tradition were recodified, could be viewed as an administratively enforced violence which works against the rebuilding of women's bodily integrity. It is a form of violence in which administrative means are used to delegitimise claims to citizenship (Beaugrand 2011). Indeed, one of the main reasons for women being unable to pursue a criminal case is that local authorities still do not consider domestic violence as a criminal matter but rather a private dispute to be resolved within the family. Bridging analytical divides between direct and structural violence, James Tyner's (2014, p. 70) work on the violence of the Khmer Rouge period has drawn on the notion of administrative violence to demonstrate relatedly how 'law-making and law-preserving forms of violence intersect with the imposition of structures of violence' (see Benjamin 1921 for formative work). This double instantiation of violence pursued through the judicial power-structure of DV law is demonstrated in Article 1:

> This law has the objective to prevent domestic violence, protect the victims and strengthen the culture of non-violence and the harmony within the households in society in the Kingdom of Cambodia. This law is in the purpose to establish a legal mechanism to prevent domestic violence, protect the victims and preserve the harmony within the households in line with the Nation's good custom and tradition.

Article 1 sets out its dual purpose not only to prevent domestic violence and protect its victims, but also to consolidate the 'culture of non-violence and the harmony' of Cambodian households. DV law displays a distinct tension, between what is hallowed as a national culture and tradition of harmonious households, set against a law that has arisen conversely from the disharmonious realities that many women face (Brickell 2015). Permeating the law, harmony within the nation and its households is constructed as a pre-existing state of affairs that requires preservation rather than adaptation for the collective good of 'custom and tradition'. As Minister of Women's Affairs, H.E. Dr Ing Kantha Phavi reiterates in her preface to the law's Explanatory Notes, 'It is my strong hope that the Cambodian legislature's key message that "victims shall be protected outside and inside the house" will assist Cambodian families in their attempts to promote the peaceful tradition of the country' (RGC 2007c, np). Moreover, Article 33 of DV law discloses that the state is responsible for education, dissemination, and training 'especially on the responsibilities within the households and respecting the rights of each other in order to promote the value of Khmer families, morality, good manners, ways of living, ways of *preserving* and educating the households' (my emphasis).

Here then the administrative apparatus of DV law promotes the value laden in stoicism and endurance to promote and maximise certain forms of life and ways of being amongst its female citizenry. In the interests of the public good, of being a good citizen, the onus is on DV victims as obligation-centred rather that rights-based citizens to keep within the civil and moral parameters that the Cambodian government sets out for them. These insights have synergies with Ayona Datta's (2016b, p. 168) findings in Delhi slums where it was perceived that 'separation from the family as a result of domestic violence was "immoral", and the use of law was ... violating the moral geographies of home, family and community'. She continues, the 'insertion of law in the home struck at the heart of the "legitimate" location of women – if families broke down, it dislocated women from their "rightful" places in the home'.

That DV law was cast in some governmental quarters as a precursor to intolerable marital and wider societal breakdown in Cambodia is evidenced in its slow and protracted journey. Two senior parliamentarians accused the Ministry of Women's Affairs of trying to incite a 'social revolution' (Rith 2005, np). In respect to the draft law, they 'balked at the proposed penalties, and they stated concerns that the draft law would negatively affect traditional Cambodian family life' (McGarvey 2005, np). When it came before parliament, 'legislators erupted in anger' as it was not something the state had the right to adjudicate and as a result 'the bill died' (Brinkley 2011, p. 233). The concern articulated was that DV law had the potential to destroy the intimate security of the family and wider society if women's rights were indulged. Privacy of familial troubles, and family stability, however much a fallacy in some cases, took political priority. The inherent variability of this logic is revealed by looking across the domestic violence and forced eviction case studies. While BKL activists' marriages stand in the way of capitalist accumulation and are made discardable (see later and Chapter 6 for a full discussion), the unassailability of marriage was insisted upon in situations of domestic violence. As Chanlina, a gender specialist in the NGO sector, told me:

> They [the government] said that we had copied from outside and said that if our proposed law was passed, it will damage our society and I remember someone accused us that we were a revolution. Yes revolution! At, at that time we were so scared of the word 'revolution' … because we had just come back from the Pol Pot regime. And the word 'revolution' is very … [tails off nervously],

Chanlina's interview shows awareness of governmental anxieties over 'outside' intervention in family life deemed incompatible with Cambodian society. Chanlina's point is that the charges of 'revolution', wielded by government officials against the NGO community advocating for a DV law, have precedence with the revolution of Kampuchea similarly predicated on the rejection of foreign influences, albeit in more extreme and complete terms. As Evan Gottesman (2003, p. 28) explains, the 'Khmer Rouge considered foreignness of any kind a threat'. The word 'revolution' is therefore particularly loaded and feared in a Cambodian context given the devastation of the Pol Pot regime. 'Revolution' as linked to DV law is suggestive of revolt and an overturning too eventful and subversive to be sanctioned.

The World Bank (2012, pp. 335–336) also notes this denouncement of the draft law for being viewed as 'antagonistic to Khmer culture'. It cites a parliamentarian's interview in a background paper (Frieson 2011) to substantiate this problematic view of the draft law, which would provide:

> … women with too many freedoms and rights, which will cause them to be so happy with their freedom that they do not respect ancient Cambodian customs …. A cake cannot be bigger than the cake pan.

In the well-known Cambodian love story *Tum Teav* the proverb 'the cake should never be bigger than the basket' features and has some echoes in the parliamentarian's phraseology. In *Tum Teav* the point being made is that a daughter (cake) should not defy her mother (the basket). The allusion in the case of DV law is that women's rights (cake) should not defy or expand beyond those of society (the pan).[3] In other words, an individual victim's needs should be subordinate to those of the national collective. As Judith Butler (2016, p. 14) writes poignantly in a connected regard, 'freedom can be exercised only if there is enough support for the exercise of freedom, a material condition that

enters into the act that makes it possible'. Instead the harmonisation of marital discord took precedence over the freedom and harmony that could come
from challenging or escaping a violent relationship. During the first debate in
the National Assembly, it was reported that Monh Saphan, of the Royalist
Party Funcinpec, said to the room: 'If the wives gain the right to report their
husbands to authorities, they are no longer wives' (cited in Müller 2010, np).[4]
The inference being made is that if women do not obey their husbands, as per
traditional ideals, then they can no longer be considered wives. They are aber-
rations of Khmer womenhood.

In essence then, DV law represented an uninvited space of intervention to
male parliamentarians who stymied its passing. Two interviewees elaborated on
this:

> In the first draft we enshrined some articles related to punishment and we submitted
> this to the National Assembly .... I don't want to blame anyone, but some Excellences,
> *especially His Excellences* said 'If this law is enacted, there will be many widows in
> Cambodia' (Interview with MOWA representative, January 2014, my emphasis in
> text).

> We observed that the male legislators did not want to attend the meeting or discus-
> sions on the draft law organized by NGOs .... I still remember Mrs. Kek Galabru[5]
> was doing a lot of hard work to lobby those male legislators. Finally, we had about
> 20–30 men who attended a meeting with us ...I still remember that there were so
> many versions of the draft DV law before 2001. I mean before 2001 there were many
> efforts on the draft DV law but it was not successful because the parliament mem-
> bers said the law was against Khmer tradition and it was controversial. There were
> some NGO members working and lobbying the parliamentarians behind the scenes
> but it was still not a success. I think it was pressure from the international community
> that finally did it (interview with Kravann, NGO gender specialist involved in DV
> law design, August 2013).

In all likelihood because of the elite male concerns evidenced in these two interview excerpts, the Explanatory Notes (RGC 2007c, p. 41) to DV law reads that it
'acknowledges and tries to limit the number of divorces'. In a rapidly globalizing
country embracing market-driven capitalist growth, the timing of these governmental interventions and clarifications is not incidental. It evidences, once again,
the home as battleground for the governmental right to manage and dominate the
home in changing times. The path of DV law reflects 'an intensification of "biopolitics" where national security has become increasingly dependent upon the
government of biological "life" in the home which, in turn, relies on the assumption
of stability provided by women giving time to others there' (Diprose 2009, p. 75).
The National Committee for Upholding Cambodian Social Morality, Women's
and Khmer Family Values (NCSWF) is an apposite example of this feminisation of
responsibility. Comprised of the Minister of Women's Affairs with members from

other relevant ministries and institutions, its aim is 'reducing anti-social behavior and increasing mutual respect, particularly for women and girls, ensuring harmony and happiness within the family and in the society as a whole' (Ministry of Women's Affairs (MOWA) 2014a, p. 8). The committee has shown regular concern for what they view as the menacing spectre of Western-style modernisation and the need to protect women and national culture through observance to traditional gender norms (Pen 2015).

State rhetoric in Cambodia is replete with the defence of culture, custom, and tradition and is pitted against what are construed as destructive influences of the West enacting cultural wounding on the nation. As Chanlina implied earlier, DV law was being framed by male parliamentarians who rebuked it as a damaging idea copied 'from outside'. It was considered an assault on the sovereignty of the Cambodian nation pressured into being passed by the 'international community' as Kravann says. DV law had thus become intimately enmeshed in 'domopolitics'; the aspiration to run the state and homeland like a protected cocoon set against a dangerous outside (Walters 2004). Echoing 'realist-inspired traditional geopolitics', this ideal of governmentality constructs the international realm as distinct from the domestic, one in which sovereign power within the nation state is absolute (Dodds 1996, p. 573). In addition to the ongoing Committee on the Elimination of Discrimination Against Women (CEDAW) monitoring the country had signed up to, the push for DV law jarred with, and impinged upon, this ideal of autonomous decision-making free from external influence. The transnational system of human rights was construed by the Cambodian government as a vertical imposition from above (this is also the case for women's forced eviction activism).

Prime Minister Hun Sen's 2011 address at the Celebration of the 100th Anniversary of the International Women's Day further elucidates this narrative, including the idea that 'fire in the house' arises in Cambodia from dangerous influences beyond the domestic boundaries of the household and nation. The speech centred on providing recommendations to the Ministry of Women's Affairs on how to best prevent and eliminate violence against women having successfully ratified DV law. The second recommendation focused on the provision of education that would instil:

> ... [the] good culture and civilization of our nation to explore the value and role of women and [the] Khmer family which is a mechanism to combat [the] inflow of negative foreign culture, eliminate bad activities among youths and children group [sic.], badly influencing our national culture, custom and norm (PrimeMinister HunSen Blogspot 2011, np).[6]

It is not uncommon that DV law is perceived as an expression of disrespect towards the family and wider Khmer culture with many local authority interviewees indicating that they were often reticent to preside over marital dissolution. While human rights are theoretically grounded in universal claims, in practice they are entrenched in 'normative imaginings of what should be' (Laliberté 2015, p. 63) or *should not* be. Offering additional analysis as to why DV law is maligned, stakeholders who worked directly with women outside of victims' immediate vicinity commonly argued that it was commune-level officials who were to blame.[7] As Mittapheap, a 31-year-old male lawyer, and Bunroeun, a 44-year-old male provincial police officer, told me (in turn):

> You have to understand the mindset of these governmental officers. They are acting as if they are the actual parents of these people. And as parents, they never want to see their children living in separation. Hence, you see that they tend to use their power to push these people to stay together … they will never report the case, thinking that it is normal for a man to abuse his wife occasionally (Mittapheap).

> There is a wrong understanding that this law is only created to divide the family … some people look down on this law … if they believe that authorities only push for a separation, then they will lose respect for that law (Bunroeun).

As Mittapheap and Bunroeun explain, a coercive paternalism can emotively work against women's interests as local authority figures effectively remove legal recourse as an option for victims who are left to manage the crisis ordinary of domestic violence. This intimate governance of the home works to mask the violence women encounter and keep crisis hidden by promoting women's altruism as the norm and ideal (see Chapter 4 on a women's altruistic load). Coercive paternalism hinders, too, women's confidence and ability to seek and receive support outside of the domestic sphere. It renders domestic violence uneventful to those in power by diffusing potentiality and returning to the 'non-emergency everyday' (Anderson et al. 2019, p. 4). As such, reconciliation is situated within what Dana Cuomo (2013, p. 856) describes more generally as 'narrow conceptions of masculinist security that often fail to address victims' multiple security needs'.

This is because DV law is perceived as an unwelcomed intrusion in the domestic sphere – a form of administrative violence in and of itself – which undermines the notion of it being an invited space of intervention. This political and moral stance has negatively influenced the ability of domestic violence advocates to press their public presence at particular moments. On Valentine's Day in 2014, for example, around 200 cyclists were stopped from biking together as part of the global mass action to end violence against women 'One Billion Rising' (Figure 5.1).[8]

Figure 5.1   Stopped 'One Billion Rising' cycling event, 2014. Source: Courtesy of GADC.

The reaction of Ros Sopheap, director of GADC, was cited that, 'the government response this morning does not illustrate commitment that the governments declare in [the] Convention on the Elimination of all Forms of Discrimination against Women (CEDAW), Universal declaration of Human Rights and [the] Cambodia constitution'.[9] As Sangiec, a *Guardian* newspaper witness, reflected at the time, GADC had 'obtained all the relevant permits for their event, and hundreds of people had gathered first thing this morning for the ride. But no sooner had we set off, we were stopped by the authorities, police and private security guards. We must have looked really menacing with our pink T-shirts and pink balloons'.[10] Whilst this disqualified event is of lesser confrontational status to those I detail in the second half of the chapter on forced eviction, it shows, once again, how women's presence and very mobility in the public sphere had been uninvited on social order grounds. The extra-domestic, and the grassroots politicisation of domestic violence in the public realm, could not be approved.

The discursive emphasis on tradition and the preservation of family unity in DV law and on-the-ground is not altogether surprising. Feminist legal scholarship has long shown how legal rules will track and reflect the dominant conceptualizations and conclusions of the majority culture (Fineman 2012). Much like the halted cycling ride, which was part of an annual global campaign, DV law speaks to the remaking of human rights in the local and the interacting process by which these are refashioned and contested. That law closely parallels the social structure reproducing gendered notions of women and social customs is plain to see in Cambodia's DV law. As a human rights worker in an NGO concluded:

> There isn't a feminist narrative behind any of the legal and policy developments relating to violence against women. So, while I think it is correct to say that the DV law have raised awareness that DV is wrong, the reasons commonly expressed for it being wrong (by members of the public, police, MOWA officials) are never that it is an expression of gender inequality or that it's an assertion of power.

Just as gender-based violence is rooted in power inequalities between men and women, the conclusion put forward here is that mainstay responses to domestic violence in Cambodia do not challenge, but in fact deny, this underlying driver of violence against women. This is despite international consensus that promoting and achieving gender equality is a critical aspect of preventing VAW. Much like Cindi Katz's (2004) notion of 'reworking', the point asserted is that Cambodia's DV law does not aim to challenge or undo existing power hierarchies but is rather a surface recalibration of them in legal discourse. As Vera Chounaird (1994, p. 434) notes, there are 'rarely unambiguous "victories" in legal change, but alternatively complex processes of appropriating law for diverse purposes and agendas. Behind the formal "universal" face of law lies a multitude of diverse subjects, causes and action plans.' As governmental charges of legislating for 'social revolution' demonstrate, Cambodia's DV law was early on in its conception viewed as an invocation of oppositional consciousness that should not be sanctioned. The contingent hospitality of law-making cannot, therefore, be understated when unravelling its incapacitated power to materially change women's lives for the better.

Rather, DV law is considered an achievement, according to Minister Ing Kantha Phavi, because it demonstrates 'that domestic violence is not a domestic issue but a social issue' (cited in Bottomley 2014, p. 7). Prime Minister Hun Sen has also made reference to violence against women as being 'not only a problem of a woman, a family or a community, but also a tragedy of the whole national socio-economy and as a global issue' (cited in PrimeMinisterHunSen Blogspot 2011, np). His speech seeks to position domestic violence as a problem and responsibility that is internationally shared rather than a 'tragedy' isolated to Cambodia. The progressive stall put in viewing domestic violence as an issue of significance beyond the home is perhaps misplaced in this instance. By up-scaling domestic violence as a global 'everywhere' problem, country responsibility and culpability is rhetorically side-stepped. This up-scaling de-territorialises ownership of domestic violence as an in-country crisis which can and should be addressed by government.

Chito, a 32-year-old male legal trainer, and Vuthanong, a 56-year-old district head of police, expressed their views on domestic violence in this regard:

> Earlier it was thought to be an issue of individuals. However, since the creation of the DV law in 2005, people have been taught that the problem goes beyond their walls. Personally, I think it affects the society more than we realise (Chito).

> People view their problems as theirs alone and ones that shouldn't be shared. This thinking is really hard for us to get rid of. It's a long-standing culture. Only now are they realising the consequences of this silence and are now becoming more vocal about their rights and freedoms (Vuthanong).

The two interviews with Chito and Vuthanong reveal how DV law is believed to have reoriented domestic violence from a private to a public matter. As Chito notes, since DV law, people recognize that 'the problem goes beyond their walls'. Vuthanong provides a more transformative narrative, arguing that despite a 'long-standing culture' of silence, women are asserting their claim rights. The DV law research found that the majority of professional stakeholders did feel that DV law had reframed domestic violence as a public not just private issue. The narratives of the two men combined with the politicians who preceded them, speak to efforts to bring the issue of domestic violence into the public domain. While few domestic violence survivors felt that DV law and their knowledge of it had improved the way they felt about their situation or ability to act, Ny's potential for a new legal self arose through listening to the radio:

> Normally when women don't go out much, they have ways of finding out new things. I was the same way. I never knew about these things [DV law] until my husband mistreated me. As a result, I went around and got information that would assist me in my fight to get justice from sources such as the radio and newspaper. When I kept on hearing the broadcasts, I started to realize that it was about my problem. Earlier I was ignorant. Now it is different. I know I don't have to put up with his anger, that I am not worthless … my health is more important than what people think … the government need to broadcast it over and over again so that people are familiar with it. Like in my case, it took a long time for the information to seep into my consciousness and makes me want to know more.

Ny had described to me her life as 'bleak' until a local woman who had herself encountered DV suggested listening to the radio. This gave Ny a new 'sense of direction'. While she was unable to act on this using DV law, she filed for divorce in 2012 marking a watershed moment in her once violent marriage. Sally Engle Merry's (2006, p.188) observation that the relationship between legal reform and the transformation of women's lives is ambiguous is apt in Ny's case; 'offering a new legal self – protected from violence by men but providing in practice a protection never fully guaranteed or experienced'. In January 2018, however, as democratic space for independent media in Cambodia shrunk rapidly, the operator of an FM radio frequency was warned to stop offering airtime to the Women's Media Centre (WMC) which Ny had been listening to. As an employee commented, over the 19 years of operation, 'We became a vocal voice for women. We have produced many programs to empower women in society and produce programs to end domestic violence' (cited in Meta and Kijewski 2018). The closing democratic space of Cambodia means that the issue of domestic violence risks not being heard to the same extent in the future if independent media is suppressed.

On another pessimistic note, it is important to indicate that the formation of rights-based identities do not come about through the ratification of new laws alone. This is especially so in a country where the very meaning and status of law needs to be factored into any redress of ongoing problems. Four different interviewees shared their views on this:

> Laws are made to have positive impacts. However, the practice of law by our citizens is still very weak. I have noticed though that it isn't just citizens who don't obey law, even public officers including me do not fully respect them … this is the culture of the Khmer people (Roth, 60-year-old male village head).

> Law in Cambodia is like a luxury car manufacturer producing upscale products for an impoverished market without doing any market research beforehand (Norin, 33-year-old male legal clerk).

> Everyone knows the traffic law but not everyone wears a helmet when they are out driving (Meng-Hour, 45-year-old male lawyer).

> Most people just plain do not care about the law. They think that they are above the law. Also, some people have a different understanding of the law. They view the law as something intrusive rather than constructive to the society as a whole (Sobin, 39-year-old female legal clerk).

Across the interviews I conducted there is a pervading sense of resignation to the impotency of law as a vehicle for rights-claiming in Cambodia. That many couples do not register their marriages or legally divorce even if they were registered (see Chapter 4) confirms this further. The excerpts presented here speak to the varied reasons why law is not generally viewed as being appropriate culturally and/or capable of protecting domestic violence victims and the general populace. In a mocking, rather exasperated tone, Norin and Meng-Hour question the assumption that law was ever, or could ever be, widely accepted or used. As Norin alludes in relation to DV law, 'market research' had not been undertaken to assess whether people would be receptive. Meng-Hour's example of the poor uptake of traffic law is a commonplace analogy. As Sebastian Strangio (2016, p. 166) elaborates in *Hun Sen's Cambodia*, the country's 'social problems can be seen in microcosm on its roads. Like the country as a whole, the country's road traffic operates less by the existing laws (which are ignored by just about everybody) than by a loose convention: small gives way to large, or suffers the consequences.' This observation connects to Sobin's point that some people 'think that they are above the law'; that hierarchy and patronage trump equality of rights and access to justice. She also echoes the sentiment of other stakeholders, that law is not always considered a positive panacea in a country where the legal framework is discretionary, there is no effective judiciary or effective rule of law (ICNL 2016). These are important overarching political economy dynamics, which are elaborated upon in the next section.

Precarious Economics of Help-Seeking

Combined with the control-based nature of domestic violence, agency to assert legal rights is constrained by an interlocking set of impediments (Nussbaum 2005; Stubbs 2002), which relate to gender-based violence as an inculcation of, and tolerance towards, gender inequality. This includes the administrative violence identified in the prior section. As Martha Nussbaum (2005, p. 175) asserts, securing a right – such as a woman's life free from violence – 'requires making the person really capable of choosing that function'. Equality of access to entitlements are in actuality differentiated according to gender, race, and ethnicity, which problematises formalistic assumptions of citizens constituting a single, all-rights-bearing entity (Lister 1997; Miraftab 2009). Domestic violence is a fundamental violation of women's basic human rights to life and bodily integrity. It is also a barrier to women's full participation in all spheres of life.

The precarious economics of women's households I introduced in Chapter 4, and their positions in broader society, are major brakes on women's agency to pursue legal action in Cambodia, for example. To demonstrate this, I focus here first on women's financial dependence on spouses and associated norms of female dependency that are mobilised by some stakeholders. I turn second to widely shared concerns over corruption, which contributes to secondary victimisation and the escalation of precarity still further. These are structural, social, and economic inequalities that (try to) keep women stuck in violent relationships and homes. They are 'hierarchical structures that entrap many women into potentially violent environments at home and work' (True 2012, p. 31).

First, then, there are strong economic reasons for remaining in an abusive relationship and foregoing rights in a context where men's predominant income earning remains the norm, especially in rural sites. In such circumstances, 'choice' is conditioned by the lack of alternatives that women have available. As a result of these factors, legal recourse is often foregone. The interview excerpts of Leakthina, Bourey, Yann, and Nuon are representative of the majority stakeholder viewpoints on this matter:

> We discussed the issue with the court and came to the unanimous decision that he [the perpetrator] should be released due to the simple fact that his family needed him to provide for the family ... sometimes it is hard to do our job because we have

to grapple between right and wrong and the ever-grey area. Keeping her husband in jail is the socially right thing to do. But it would inevitably hurt the whole family (Leakthina, 53-year-old female police officer).

It is very difficult to fully implement this kind of law in developing nations such as Cambodia. It is more effective in a developed country due to the availability of jobs and skills for women. Khmer women are mostly dependent on their spouses for financial support. They are bound by this ugly truth. In developed nations, women have more financial freedom; and therefore, social freedom and mobility. However, with more freedom, there also comes more broken families … more marriages in developed nations break apart compared to a place like Cambodia. So, I guess, it is a give and take scenario … while I applaud the creation of such laws, I am having a hard time embracing it wholeheartedly because it lacks so much in terms of providing options for the victims and their families. The law fails to provide better alternatives for the victims. Let's also say that I sentence the perpetrator to jail term. What I do is essentially taking away the only provider of that family. If you were the wife or children of the household, you would not have any other choice than to spring the culprit out of jail so that he can go out and earn a living to provide for the entire family (Bourey, 50-year-old male court judge).

Small-scale physical violence does not get brought up in court. The reason being that the local authority equates them to nothing more than household squabbles, which are noncriminal in nature. A slap here and there is just that, a slap. Also, we have been dealing with a lot of cases where the victims themselves refuse to press charges, arguing that their spouses, however violent, still provide them the type of financial stability they need to sustain the household. This is pretty much the majority of [the] issue we face here (Yann, 31-year old male NGO legal trainer).

Normally before the government creates a law, they should create the implementation strategy and budget but here, they only create the law. That is why there are too many problems in the real practice. It's impossible (Nuon, female NGO worker, March 2014).

As the stakeholders each remark, although DV law is a positive step forward, its application in their professional practice is hampered by an array of challenging issues related to the extra-domestic conditions in which women's rights-claims are embedded. A long-term specialist in the policing of DV, Leakthina highlights the moral quandaries she faces on a daily basis as she makes judgements that straddle 'right and wrong and the ever-grey area'. An example of this are decisions taken to release perpetrators by professionals operating within a constrained decision-making environment. Working in a country devoid of a welfare or legal aid system, perceived options open to both victims and professional stakeholders are constricted. In addition, as LICADHO (2017, p. 11) have identified, 'There is no state provided social security system in Cambodia and other than a few NGOs providing limited safe shelter to victims of domestic violence, there is no provision to assist women leaving a violent situation' (see also in-depth research by Graham 2019). For women who do access shelters, guarantees of safety from injury and even murder is enabled through their spatial exclusion from mainstream Cambodian society as a 'safe space' for perpetrators (Graham and Brickell 2019). Many women's financial reliance on men also means that they either cannot, or will not, choose to seek legal redress given its risks.

Again harnessing women's altruistic burden, the interviews indicate how victims' needs are balanced against those of the household as an economic unit that must remain viable and sustainable for the well-being of its younger members. As Yann comments, 'a slap here and there' is not deemed serious or exceptional enough to warrant the discontinuity of the household. In other words, domestic violence is part of the everyday panorama of the crisis ordinary in which a slap is a self-evidential and careless act devoid of greater meaning or eventful status. 'A give and take scenario' similarly dominates Bourey's narrative who comments that with freedom invariably comes broken family structures more common to the developed world than Cambodia. Victims' exit options from violence are hindered by the 'ugly truth' (as Bourey phrases it) of financial dependency as a structural inequality that holds women in place. Furthermore, as Nuon notes, the budgetary resources were not put in place to facilitate the proper implementation of DV law and provide the support victims require to live independently from their husbands.

Domestic violence victims also spoke of precarious economics, of not having well-paid or secure enough employment. Being alone, many felt, jeopardised the financial means to run an independent household and ensure that basic needs were met, from food to school uniform and fees. These findings speak, once more, to the significance of political economy dynamics for situating women's experiences. Precarious work and precarious life are therefore intimately connected in a web of structural violence. As a result, some women emphasised the importance of ensuring their daughters were well-educated to avoid a similar fate to their own. Kesor shared that:

> I think on paper men and women have equal rights. You can say to women you have equal rights, but if men do not cooperate then do these have any meaning? Women's financial freedom is serially taken away .... I witness it, it's a big problem ... husbands hand over very little money to their wives and use it instead to get drunk. You can talk about equal rights, but this is not equal .... I am trying to encourage my daughter to do well at school so that she won't be stuck with anything or anyone who could potentially tie her down, including a failed romantic love. I want to make sure she looks beyond her comfort zone and sees what the outside world can offer. I don't want my daughter to have to put up with the crap I have dealt with for the past 10 years .... I don't want women to be men's punching bag anymore.

Kesor believes that her daughter's financial independence would remove the impediments to escaping violence which she herself has endured for a decade. She places onus on education as the vehicle that will hopefully afford her daughter

the ability to leave the 'comfort zone' and avoid being confined to a violent relationship. A female director of an NGO in Cambodia also saw this connection:

> It is still believed that society stereotypes a divorced woman as a bad woman. Why
> did her husband accept her want for a divorce? A good woman always stays with her
> husband forever. I have seen many cases of ordinary women who cannot be
> independent in terms of earning an income in support of the family. If a woman can
> earn her own income, she would not so easily become a vulnerable victim of
> domestic violence. She would have more choices to decide her own life. For example,
> she can talk to her husband by saying 'if you continue to commit domestic violence,
> I can go my own way in life'.

Yet even when a woman might have enough economic resources to pursue legal
action and/or leave a violent marriage, some stakeholder interviewees believed that
the prevalence of gender norms continue to have deleterious consequences for DV
victims. A director of another Cambodian gender-focused NGO even contended
that discourses of dependency were being mobilised to keep the status quo in situations where women's income earning was strong and would afford the funds
necessary to take formal legal action.

> Social attitudes about gender roles play an important role in women's ability to
> access justice regardless of the economic reality. For example, many women actually
> are the breadwinners in their families, but because their community, including their
> families and law enforcement officers, believe that women should be supported by
> men, those women are unable to get assistance when they try to leave.

Mooted here are generalisations and presumptions of women's dependence
that are being used to uphold the moral hegemony of the family and wives as their
ever-present yet male-dependent lynchpin. As Srimati Basu (2012, p. 489) writes
in the case of India, legal deliberations in court 'exist within a cultural and legal
framework of women's economic dependence on marriage for food and shelter
and male control of women's decisions and movements, and in which domestic
violence is seen as a personality-related aberration and wifehood and motherhood
are associated with tolerance and sacrifice'. Indeed, the myriad excerpts in the
preceding discussion of precarious economics – both actual and strategically
construed – speak to the operation of law as much as a mirror of society than as
a catalyst of enduring change.[11]

The capacity for law to protect domestic violence victims is elided still further
by secondary victimisation and the cascade of precarity from corruption con-
fronted when help-seeking. Secondary victimisation encompasses 'injustices that
occur to victims after a trauma' (Hattendorf and Tollerud 1997, p. 17), and which
betray a victim's expectation that 'she will be provided with belief, validation, and
protection when she instead encounters victim-blaming attitudes, or her victimization is ignored or minimized' (Laing 2017, p. 1316). Kesor's experiences again
witness this:

Unless the authority takes the law seriously, nothing will change. I mean that if he becomes aggressive and his behaviour warrants arrest, he should be arrested and sent to jail for a determined amount of time. The authority should show no favour and take no bribes. However, it's not like that. You know when my husband is arrested, he just says that he only needs to bring this much money with him, he'll be out soon. He even tells me that there is a kind of jail that lets you party and bring girls in too.

Most of the women interviewed in the DV law project had experienced revictimising dynamics in the Cambodian legal system that echoed those that they had experienced from their husband, namely silencing, coercive control, and economic violence. As Lila Abu-Lughod (1990, p. 52) remarks, 'resisting at one level may catch people up at other levels'. In Kesor's case she is taunted with impunity by her husband who brags that he can pay off jail wardens with ease. CEDAW (2013, p. 3) have, relatedly, called for action in Cambodia to 'investigate and prosecute allegations of corruption in the administration of justice, and, where applicable, punish the perpetrators'. It remains the case that legal recourse remains something that is often discouraged because of the dangers of corruption that victims may face, including the extortion of informal fees.[12] DV law therefore offers up the opportunity for necrocapitalism, indirectly benefitting from women's exposure to suffering, and directly from administrative and economic violence in the form of corruption. Three stakeholders explained more:

It's [DV law] been a sort of rollercoaster ride, to be honest with you. The issue is not the laws but rather the implementation of such laws. I don't want to go deeper because I think that it will lead to other problems ... but as you know, most of our officials are not keen on obey[ing] the laws. Furthermore, those who are tasked with enforcing such laws do so only [to] benefit themselves. When no such profit can be extracted, they choose to ignore them and let the people suffer (Sarom, 53-year-old female deputy village head).

We are attempting to work with those victims who distrust the authority because of the level of corruption. They stand to experience the most abuse because the perpetrators realize that the law does not prosecute them ... the level of corruption here is incredible. If we are vocal about this type of corruption, we stand to isolate ourselves from their assistance in the future. It is hard for us to say because these officials are so underpaid that it is almost comical .... I completely understand that these officials do not make a lot of money doing what they do. However, to ask these villagers for hundreds of thousand riels is not a viable practice.[13] The rule in the book is completely different than what is really happening when DV occurs. The authority tends to neglect its obligations, saying that those issues are beyond its capacity. The reasons being that those people have not been bribed. To get any legal paperwork such as a marriage license, birth certificate, and stuff like that always costs people money. People have become understandably weary of the law (Pheakkley, 31-year-old male NGO worker).

You have to look at a bigger picture here. Let's say that we arrest her husband when he beats her and send him to jail for his crime … when she needs to bail him out, the police officer responsible for the arrest will hit her with some bogus fine and she will have to pay him in order to get her husband out. You see, either way, she will be the only [one] losing out in this whole deal (Chantrea, 65-year-old female deputy district leader).

The trio of excerpts displays a reticence to talk about corruption, although evident in plain sight and roundly accepted as a characteristic feature of Cambodian daily life. Many district and sub-district stakeholders also had networks and hierarchies of patronage and power to be cognisant of in their daily work interactions to ensure their own professional survival. Given the mismatch between legal rights and on-the-ground routine, rather than assuming an absolute anti-violence stance, many legal practitioners constructed women's survival within an existing marriage. Women were in danger, Chantrea argued, of experiencing a lose–lose situation by taking on the risks of legal action and then needing to renege on them given the precarious economics of their family's survival combined with corruption. As a female opposition politician asserts, corruption and impunity goes to the heart of problems surrounding the inefficacy of law to aid DV victims:

The key to domestic violence is that if we are ever going to solve it, if you don't have a court system, a just court system, it doesn't matter what kind of law you have … let's say that we have a law that's, uh, a standard, of international standard … you can't trust the law, you can't trust the authority, you can't trust the police, the law is corrupt, the court is corrupt so how are you going to, how are you going to, um, how are you going to target them … so, domestic violence will always be an issue as long as there is no judicial reform … let alone not having a decent domestic violence law! [laughs]

Stories like Tola's testify to the point being made here in graphic terms. Tola was denied legal help altogether because the family were known to vote for the (now dissolved) CNRP opposition party.[14] According to Caroline Hughes and Netra Eng (2019, p. 372) 'maintaining positive relationships with local authorities remains an important survival tactic and local authorities continue to routinely exclude households known to support opposition parties …. This encourages strongly deferential attitudes and a deep fear of falling into disfavour with local chiefs' (see also Schoenberger and Beban 2018 on these fears).

When interviewed, 32-year-old Tola had been married twice. In her second marriage to Akra she had encountered years of his infidelity, alcoholism, and severe physical violence during and after two pregnancies. She believes that for two or three days per month she avoided being pelted with flying china, a television, or fan. Tola describes years of nausea in which she lost a significant amount of weight (much like Orm did in Chapter 1). She also explained escaping to her sister's house only to find her husband going to the police and maliciously accusing her brother-in-law of beating *him* up. She now lives with her mother and eight other household members including siblings. Tola explained what happened:

> I went to complain to the village and police office again but they asked how much I was willing to pay them. We must have money, otherwise we still lose the case even if we are beaten many times. Also, a long time ago, my father voted for a political party, a different party from the village office here, which was against them … our party was the Sam Rainsy party,[15] so they said they wouldn't file a complaint.

The misconduct of local authorities in Tola's case not only reflects the politicisation of their respective positions but also the hierarchies of power, connections, and status in operation. As Fabienne Luco (2002, p. 18) explains, the 'village chief is now the local representative of the main political party, a situation that creates suspicion and divides villagers'. Having been denied help, the physical violence continued unabated given the administrative violence of local leaders who acted with impunity. Returning from her mother's house one night, Tola was electrocuted as she entered their home. Akra (an electrician) had intentionally rigged up an exposed wire against the metal door. Even having reported this violent escalation, Akra was only instructed by the police to change his behaviour. Tola meanwhile was told to change her 'attitude'. Desperate, Tola negotiated a divorce as the 'best solution' to her protracted problems but five months on decided to reunite with him under the financial stress of debt incurred to secretly 'pay-off' the court US$1200 to expedite the divorce and secure child custody. Again, the violence reignited. Akra pushed her into the fire and covered her with a blanket and mosquito net. As she talked, Tola proceeded to show her arm and leg burns in the interview.

> Some neighbours saw it but they couldn't help me. They weren't brave enough to help. They pretended not to see me because they thought it was my family issue. I didn't know who to ask help from. I wanted to hang myself but I was thinking about my children. I went to hospital to ask for some cream because I was in intense pain. I told them my husband burnt me. They advised me to file a complaint at the village or police office. And otherwise to go to an NGO for help … of course this is what I did as the local authorities would have just asked for money. The NGO took a photo of me and asked that I go to the commune office telling me I shouldn't skip them. So I went there and the police came to my house. They said because we were divorced Akra needed to stay away from the house. He agreed to leave but failed to do so for three days. I didn't stay there because I was afraid he would choke me to death. I went to complain to the police office but they did nothing.

Tola's interview has a close correspondence with those I presented in Chapter 4, in the respect that the seriousness and intractability of the violence she was encountering meant she had considered suicide. Her interview also indicates how domestic violence remains, even in acute moments of violence, considered a 'family issue', which neighbours are unwilling to intervene in. As Thavy Thon (2017, p. 156) notes in her self-published story of navigating gender norms in contemporary Cambodian society, domestic violence was an 'open secret' in the village she grew up in.Yet despite the abuse being heard in her closest neighbour's house, 'no one did anything to stop it'. For the majority of domestic violence victims in my own research, intimacies of care shown by neighbours were also unforthcoming. Despite being physically close in terms of distance, for many women the lack of support had a privatising effect on their grief.

Rather than retreat from seeking help, however, in Tola's case she invested faith in an NGO rather than the local authorities who she suspected of corruption. Tola continued to explain how her advised return by the NGO to the commune office bore these concerns out. The police again refused to arrest her husband without payment. The couple went on to break up after Tola refused to live with her mother-in-law. Since this time she has been supported by a local Christian NGO who provides employment skills training for those who have suffered violence.[16] The combination of factors identified in this discussion of DV law means that for many women there are few options other than 'getting by' in the crisis ordinary despite the existence of the law for more than a decade.[17] Violence against women is, once again, normalised as a part of everyday life that should be managed rather than overturned.

Contesting Forced Eviction

If DV law sits ambiguously as a sanctioned yet uninvited space of intervention, then forced eviction activism is unequivocally an unsanctioned and unwelcomed rights mobilization from a state perspective. While women are too often represented 'as passive objects of impersonal and unstoppable economic forces' (Pratt and Rosner 2012, p. 3), of which forced eviction is one outcome, BKL women have contested the abjection of home and the Cambodian government's malevolent logic of national development. Through invented and eventful spaces of bodily presence in the public domain, they have railed against the (social) death that forced eviction can ferment.

In this second half of the chapter, I focus on women's legitimation of claims to dwell. In circumstances of forced eviction and fighting for these rights, the 'protagonists of this citizenship drama use nonformalized channels, create new spaces of citizenship, and improvise and invent innovative practices' (Miraftab and Wills 2005, pp. 201–202; see also Cornwall 2002; Holston 2008; Velásquez Atehortúa 2014). I explore this drama in two sections. The first looks to the whys and wherefores of women's leading role in contesting the BKL evictions. In the second I turn to the improvised and innovative practices with which they used to do this. To flesh these out, across both sections I draw on the narratives of BKL women (Soy Kolap, Srei Pov, Srei Leap, and Heng Mom) and two husbands of group members (Visna and Phala).

Defending Home: Women's Leadership Against Forced Eviction

Since 2008 BKL women as wives, mothers, and committed activists have steered the campaign of non-violent social action against forced eviction and in so doing have shown an alternative to those invited spaces of citizenship organised and sanctioned by the government via DV law. According to BKL interviews, the group emerged as a result of women's strong pre-existing networks within the community, which solidified in the face of shared trepidation. As Judith Butler (2004, p. 22) explains, 'Many people think that grief is privatizing, that it returns us a solitary situation and is, in that sense, depoliticizing. But I think it furnishes a sense of political community.' Indeed, 'grief activism' can 'foster relationalities and communities in opposition to a politics of division, abandonment and necropolitics' (Stierl 2016, p. 174). Unlike the isolating experiences of Tola described earlier, the relational ties between women with shared identities, under similar burdens, led to their cohesion and collective defence of homes.

BKL women's activism illustrates the importance of taking home seriously as a 'territorial core' (Porteous, 1976, p. 385), which women defend perhaps more than any other level of fixed physical space. This is a territory that domestic violence victims are often sent back to by authorities, but which forced eviction victims are removed from. As I have argued in the book so far, forced eviction is a form of gender-based violence that is predominantly directed at, and negatively experienced, by women because of their foremost homemaking roles and identities in Cambodia. These, as Silvia Federici (2018b) posits, stand in the way of capitalist accumulation in the case of forced eviction.

Comprehending how BKL women have challenged this, means being attentive to the historical backdrop of genocide that women emphasized to draw attention to the gravity of evictions they contest today. Soy Kolap explained:

> We do not want Pol Pot to happen again … the fact that no one stood up and protested resulted in massive killing in our country. We were too quiet during Pol Pot Regime because many of us were illiterate, and our parents wanted daughters to keep silent and did not send us to school; instead we were taught to cook. During Pol Pot I was so afraid and worried that my parents would be killed at any time. I couldn't do anything besides kneeling down and crying. Now I am educated and understand my rights and laws I can help people who have similar problems to me. I am very grateful to civil society and NGOs who have trained me. I am more thoughtful and braver too. I cannot let this tragedy happen again; therefore, we protest for our rights.

Fifty-year-old Soy Kolap draws on the historical antecedent of genocide to emphasise the necessity for social action against forced eviction. Here she re-evokes and re-connects crisis temporalities stretching back to the past and forward into the present. While crisis is often used 'to signal a distinct temporality of urgency and exceptionality that works to overlook the need for more longitudinal and in-depth attention to particular circumstances' (Ramsay 2019b, p. 2), Soy

Kolap horizon scans to problematise the chronology of post-conflict peace so often told (see Chapter 3). Her interview speaks to a notion of crisis temporalities that emphasise individual and collective apprehension of emergencies that can be discrete notions of time and/or enduring possibilities (Itagaki 2013).

Soy Kolap makes her case against forced eviction and women's need to mobilise to protect their futures given the multiple limitations of doing so during the Pol Pot era. Providing a gendered analysis, she highlights the norm constraints and lack of education prior to the Khmer Rouge that she felt quashed any potential resistance to the regime by women and meant only being silent, kneeling down, and crying. During the Sangkum period (1955–1970), 'education was blamed with inverting the "traditional" position of women in society' (Jacobsen 2003, p. 204). More specifically, this manifested itself in a concern shown by parents that if 'their daughter could read and write love letters before they had the opportunity to arrange a marriage', they 'could be used to prove that the girl was not a virgin', which would degrade their social prestige and the number of presents received (Jacobsen 2003, p. 205). Soy Kolap indicates the educational progress and civil society training since that has improved women's knowledge of their rights and has emboldened her rights-claiming. In these senses, education and greater gender equality represent tools of change against gender-based violence. Education is perceived to have the potential to challenge existing hierarchies of power; it therefore renders the educated as 'dangerous citizens'.[18] Interviewees Soy Kolap and Srei Pov continued this line of thought (in turn):

> Frankly, when I was a child, I was taught to obey a husband and keep the house. That protesting had nothing to do with women. However, we have modernized over time, and nowadays, women and men are equal. If women are equal to men, then they cannot sit down and wait for their houses to be ruined. If women have equal rights, we must claim our houses back … (Soy Kolap).

> Now if women sat still and did protest we won't have a house to live in. I would like to respect the *Chbab Srei*, uphold my responsibility as a wife and housewife, yet at the same time, I have to protest. I believe that it is not wrong to do it or to disrespect the *Chbab Srei*. I am positive that if women's lives in the past were infringed so badly like us today, then they would also protest the same way we do (Srei Pov).

To elicit these responses, I had asked Soy Kolap and Srei Pov if they felt women's public protest was a modern idea. According with Soy Kolap's recollection of her childhood, there was, before the 1970s, a 'close correspondence' between the ideals articulated in the *Chbap Srei*, other literary sources, and the ways women talked about their lives in Cambodia (Ledgerwood 1996).[19] For a woman who aspired to be a 'successful Khmer woman', valued by society, it was important to be 'virtuous' and conform to the conventions instilled by the *Chbap Srei* (Ledgerwood 1990, p. 24). In the 1950s and 1960s, for example, nationalists had paraded the *Chbap Srei* as a form of resistance to French colonial influence (Jacobsen 2008) and in the post-Khmer Rouge period, the reassertion of 'tradition' through such conservative literature was also once again evident (see Jacobsen 2008). Soy Kolap and Srei Pov complicate the enactment of 'masculinized memory' (Enloe 1990, p. 44) of Cambodian women as passive and inferior to men, however, by using the history of the Khmer Rouge as their own strategic resource to 'invent' their provocations for emboldened female protest. Complicating any neat distinction between continuity and rupture in contemporary Cambodia, they argue that being respectful of the *Chbap Srei* does not preclude their activism. The rules demarcate that:

> The bad kind of woman kicks loudly
> When she walks very loudly they consider her step like a lightening sound so that her sampot [sarong skirt] is torn apart
> She walks very loudly
> So that the house trembles
> The bad woman sees something on the ground and then she moves forward without picking it up
> In the future if she cannot get organised
> Then her property will be lost

This section of the *Chbap Srei* relates particularly well to the point Soy Kolap and Srei Pov are making; while women's activism may be 'loud' in its physicality and presence – and thus fails to heed guidance in the *Chbap Srei* – these actions are necessary so they do not lose their properties. The auditory analogy marks out the eventfulness of their activism such that it politically registers outside of the home (this is akin to the symbolism of the bikes, which domestic violence advocates took to and which met political resistance). The position taken is that being courteous to gendered norms and ideals in Cambodian society is not mutually exclusive from their activism. In other words, their activism honours their responsibility to protect and nurture the home. To some BKL women, their activism is also conceived as a form of risk mitigation that includes saving their spousal relationships. Marital precarity and the continuity of gender norms surrounding women's domestic responsibilities were therefore germane to BKL women's rationale for defending their homes. Without homes they were likely to suffer the perceived social death of a 'widow woman' (as per Chapter 4) and connected psychological harm. Soy Kolap told me:

> … it is commonly said that women who have ownership of their houses and can decorate their houses nicely will have a longer lasting marriage. If not, husbands will always look for other girlfriends who have houses and who are capable of keeping their houses looking lovely. That's why women are well aware of their rights, and it is the grounds for women leading the protests.

The 'oppositional consciousness' that Cindi Katz (2004, p. 251) casts as core to 'resistance' is therefore more complicated than might otherwise be understood. The women couch their protection of home as an extension and elevation of their traditional responsibilities as wives and mothers to ensure family harmony and stability (something also emphasised in DV law). Therefore, while Tep Vanny has spoken publicly of the group's desire to show that 'We can do more than take our husband's clothes, wash them, and hang them' (cited in Vital Voices 2013), this does not mean their protest is to the exclusion of normative responsibilities. Rather it mobilises them for insurgent, and at the same time, conformist ends.

BKL women's narratives bear a striking comparison with that of the Mothers of the Plaza de Mayo nearly 25 years ago in Argentina. Demonstrating against the disappearance of their children who were suspected to be opponents of the ruling military regime, the Mothers continued to assert their identities as housewives and mothers through their organised activism on the streets. As Marguerite Bouvard (1994, p. 80) explains, they 'were still protector of their children, but in this distorted universe that meant entering the labyrinth of the political system instead of cooking or ironing their clothes'. BKL women spoke similarly of the transition from their self-confessed interiority as housewives 'living like frogs in a well' to a notorious public presence on the national stage.[20] With this a series of spatial reversals were put in motion with their menfolk. These reflect how gendered divisions of labour in the private sphere shape access not only to the public sphere but also to political, economic, and social rights that derive from such access (Lister 2003). Visna and Phala, husbands of BKL activists, explained more:

> Women do not forbid us from leaving the house when they are protesting. But, we understand that we are living on conflicting land and so we have a lot of enemies who want to create chaos to our area and threaten our children while their parents are away. Women protest outside, and men protect security back at home (Visna).

> I do not mind having the roles shifted. A married couple doesn't just live together for one day, it's for the rest of their lives. Who could just stand and see our wives having problems alone! If husbands do not take on responsibilities to keep the house and look after the children, likewise sending them to school, what then? I think it is fair enough that men take these responsibilities on … but husbands can return these responsibilities back to their wives. This is why I said husbands and wives spend a lifetime together (Phala).

Many husbands drew on the trope of masculinist protection to explain their domestic guardian role, of ensuring the 'security' of their home and their children from siege while their wives were on the frontline of protest. For example, a language of intimate war laces Visna's narrative as he describes the 'enemies' that threaten their home and children. A rare piece of journalism on BKL women's partners ran with the lead 'Boeung Kak men battle on homefront', which also evokes this connection (Worrell and Chakrya 2013).[21] Others, like Phala, had taken on more responsibility for childcare and the related logistics of everyday family life in this war effort. Such suspensions were temporary rather than permanent recasting of normative Cambodian gender roles in marital life however. As Phala makes clear, the expectation is that wives will return to backstage duties once their frontline protest ends. That marriages are notionally life-long affairs means that such transitory arrangements could be accommodated *for now*.[22] The durative dimensions and complexities of crisis can therefore be told through the dynamism of marital arrangements. Indeed, this inversion of the male/female, public/private dichotomy – so core to feminist geographical critique – is imbued with a temporal contingency that spans across this book. That women's defiant tactics detailed next were not as short-lived as envisaged had intimate consequences for the sustainability of their marriages, which I turn to in Chapter 6.

Hitting a Stone with an Egg: Women's Defiant Tactics

I have been told that villagers are eggs, and those powerful are stones, that we cannot win against them. But I don't think that way. It is probably true that we are the eggs, and they are the stone[s], however, we have to clash against the stones even though we might be crushed. At least I can make those stones smell bad. It's similar to water which drips on a stone and over time makes a hole …. I am optimistic that my efforts are not valueless.

Srei Pov's reference to eggs and stones connects to several Khmer proverbs that imprint the importance of tradition and respect for the way things have always been done. 'Don't hit a stone with an egg' (*pong man kom chual neung thma*) signals the futility of trying to challenge the dominant or powerful. This is a message that can be read through the domestic violence material and the keeping of the status quo. Indeed, the proverb mirrors the 'broad legitimation of hierarchy and privilege' within Cambodian society that perpetuates acceptance of the social order and the navigation of relationships (Hughes 2003, p. 61). BKL women's activism was underpinned and animated nevertheless by anger and hope that their actions would not be futile; that sustained repeated acts over time would have attritional potency.[23] While Chapter 4 introduced the attritional warfare used to wear down the resolve of residents who refused to accept compensation and leave BKL, this example shows it being used by women as a tactic.

At the start of their protests in September 2008, gatherings outside Hun Sen's house and the putting up of 'Stop Forced Evictions' signs on BKL houses were commonplace activities. The same month, with support from the Center on Housing Rights and Evictions (COHRE) and an attorney, BKL tried to submit an application to the Phnom Penh Municipal Court for an injunction to stop pumping sand into the lake. The injunction was denied and the subsequent appeal at the Court of Appeal in December 2008 was also dismissed. Based on an interview with Natalie Bugalski, who worked for COHRE, Chi Mgbako et al. (2010, p. 53) explain how, 'Even though both parties believe litigation is unlikely to prove successful, they argue that it is their duty to challenge the court system in an attempt to show the judiciary that the residents and legal community in Cambodia will fight to protect their rights.' In other words, in the maelstrom of patronage, corruption, and impunity (noted earlier in relation to the administrative violence of DV law), it was known that the community's legal rights were unlikely to be entertained, even less upheld, but the act of trying still mattered.

Given this unreceptive national context, BKL women looked to other spaces for potential redress. On behalf of 4250 families facing forced eviction, the women spearheaded a challenge against the World Bank's collusion in their forced eviction. In April 2009, with the help of several NGOs, the World Bank Inspection Panel was forced to investigate their claims that its Land Management and Administration Project (LMAP) had adversely affected residents by arbitrarily excluding BKL and thus failing to uphold systematic land registration. In March 2011 the panel ruled that the World Bank had not only 'breached its operational policies' but 'found that these failures contributed to the forced eviction of BKL residents, who were unfairly denied the right to register their land through LMAP before the Government leased the area to a private developer, ultimately leading to their involuntary resettlement and forced eviction' (United Nations General Assembly 2012, p. 65). The result was hailed in the national and international media as a rare victory. Following these findings, in August 2011 the World Bank reportedly attempted to remedy the breaches under the LMAP, but the government ended its project unwilling to cooperate with the Bank's suggested remedial actions (United Nations General Assembly 2012). The World Bank subsequently suspended funding for all new projects to Cambodia (see Chapter 6 for further information on what has happened since). As this example testifies, the strategies of activists like these 'demonstrate that the mobilization of scale has proved an effective political strategy' (Mountz and Hyndman 2006, p. 459) through the holding to account of an international multilateral organization and, by extension, its lending to the Cambodian state.

Several years into the dispute, Srei Pov and Tep Vanny took a twin lead in representing the group at the multiple press releases and protests held. In 2011 differences of allegiance amongst some members came to a public head one week after the World Bank's decision to freeze future lending to Cambodia was leaked to the press. The City Governor held an event at City Hall to mark the granting of a land concession to some residents still living around the lake. A total of 12.44 hectares was granted to the remaining 779 families. The municipality, however, arbitrarily excluded 96 families who could have benefitted while giving some of the land to the company instead of the families (Human Rights Watch 2015).

The film *A Cambodian Spring* (Kelly 2018) shows BKL women seated on the front row of the decorated gazebos at the event. After reminiscing about how wonderful the BKL area once was, Kep Chutema, City Governor, summates in the film:

> Now that has all come to an end. The government has approved the division of the land into two parts: 144 hectares to Shukaku Inc. for development, 12.44 hectares to the residents of BK. Tax free. [applause in the crowd] This is what we call democracy. We do not fight into the streets. As Prime Minister Hun Sen has said, again and again, it is no longer fashionable to solve our problems with violence. But let me be clear. Don't think that the World Bank will help you. They will not!

Noting the coming to the end of BKL, the City Governor uses a language of closure in his speech to impress the already-lost and futility of women's eventful protest. Albeit implicitly, he also evokes the Khmer Rouge period in his call for the halt of women's protests or what he construes as fighting on the streets. Kep Chutema then asks for group member, Ly Mom, to come forward to read a letter of appreciation to Prime Minister Hun Sen. The rest of the BKL women stand up, waving their hands, and shouting, 'Ly Mom is not our representative! Ly Mom cannot be our representative!' Through the commotion, Ly Mom continues to thank the wise leadership of Hun Sen along with his esteemed wife for 'giving his full attention to the difficulties of the Cambodian people'. Mr Kep Chutema turns to the angered women and asks if they consider Ly Mom to be their representative; a question that elicits a negative response and his scorn. After Tep Vanny gives her own short speech, the governor cautions, 'you have been given your land titles, so go mind your own business'. Applause erupts on one side of the gazebo where Ly Mom sits, while on the other, stony faces stare on. Once part of the activist group, she had 'sold out' one woman utters.

The provocation of the World Bank 'from below' and the defection I have just detailed did not alleviate the continued need for other more grounded and localised forms of resistance. BKL women persisted in pressing Prime Minister Hun Sen to intervene through further demonstrations outside his house and in Freedom Park. Engendered both in the DV law lobbying discussed earlier, and the forced eviction activism covered here, are requests made for governmental intervention to support women's rights-claiming. In the BKL case, Shukaku offices were also sought out. The women used petitions as a means to build support and staged their delivery to various countries' embassies, including those of Britain, South Korea, and the United States. UNDP were targeted to coincide with the then UN Secretary-General Ban Ki-moon's visit in October 2010.[24] These actions gained growing international attention. An ever-increasing presence on the streets of Phnom Penh, BKL residents, often over 100 in number, engaged in the blocking of key arterial roads such as Monivong Boulevard (e.g. April and

November 2011).[25] During all this time, evictions, sometimes successfully halted, nevertheless took place.[26]

Women's repeated demands for an immediate halt to the surge of water and sand into their homes continued to be denied. Following evictions on 19 September 2011, in which multiple homes were torn down (see Chapter 4), the first of many SOS actions began. In its aftermath evicted residents used their temporary shelters to put up signs sending an SOS to the public (Figure 1.1). Others meanwhile asked, 'Where is my house? Where is my right?' Since 2012 BKL women have moved from these conventional tactics to more creative interventions to exert their rights-claiming and ensure their commitment to defending their homes.

Opening the door on to women's activism and performance of citizenship meant that I entered the front courtyard and adjoining shop of Tep Vanny's home (Figure 5.2). It had become the advocacy office for their campaigning and where

Figure 5.2   Boeung Kak women's workshop, 2013. Photo: Katherine Brickell.

their creations were conceived of and fashioned. Organised from this domestic locus, *The Cambodia Daily* (Narim and Zsombor 2012, np) observes that BKL had brought 'a touch of theater in a growing list of attention-grabbing tactics' that accompanied their mainstay petitions and official complaints. They had made their plight and cause 'eventful' and challenged the norm that 'phenomena and events that are commonly viewed as public, political, global and spectacular ... have wider appeal as subjects of study than the private and apparently mundane' (Pain 2014, p. 532). They had put the intimate violences of forced eviction within an emergency register. Here, then, 'emergency becomes a resource and tactic to claim a future' by making 'attritional lethality into events that demand some form of urgent response' (Anderson et al. 2019, p. 13).

My personal correspondence with Kevin Doyle (2018), editor of the *Cambodia Daily* newspaper between 2004 and 2014 reveals more:

> While their early protests involved little but their physical presence, their later demonstrations involved directed themes with prepared props, such as designed and Photoshopped images of their imprisoned neighbours, or, for example, Styrofoam scales of justice, and what once appeared to be papier-mâché head-dress in the form of a chicken, and another time quite elaborate (model-sized) paper and wooden houses, which they attempted to carry on their shoulders to Phnom Penh Municipality offices. The use of such demonstration props made for interesting visuals for the media and ensured a greater chance of coverage – even international – of what had become routine protests. However, their most penetrating props were those that relied on traditional Cambodian motifs and superstitions, and what some refer to as black magic. They fashioned life-sized effigies, in actual clothing, of 'corrupt' government officials, and dragged them along the road with ropes tied around their necks, to the front gates of the Ministry of Justice. A ritual was then performed where salt and rice were thrown on the effigies. One such ritual featured a dead chicken's feathered carcass, splayed out using a small wooden frame. I had previously seen Cambodian troops use the same 'black magic' with chicken carcasses in their trenches facing Thai troop positions during the Preah Vihear temple conflict. While the Boeung Kak protesters' use of props demonstrated a high level of media visual awareness, and their growing demonstration sophistication, it was still their (physical and symbolic) public resistance to oppression and injustice that made the Boeung Kak protesters so powerful.

In this intimate war, BKL women's tactical use of visual rhetoric created a series of 'image events' (DeLuca 1999), which commanded media attention given the affects and controversies they provoked. Their activism revealed, but also became engrossed in, the bio-necropolitical order of Cambodia; the dragging of straw effigies and the use of black magic exemplary in this respect. 'Normally at funerals, we throw money, candy and flowers for the corpses. This time we threw chili, salt, dust and water' one resident explained (cited in Jackson and Monkolransey 2015, np).[27] Cursing rituals were acts of anger harnessed as alternative 'non-violent' means of communicating the injustice of the BKL situation, channeled through bodily practice. That these had previously been harnessed by the Cambodian military, and the political elite to gain 'magical supremacy' over the territorial dispute at Preah Vihear (Christensen 2016), suggests women's mimicry of this spirited politics in their own territorial dispute.[28] Such supernatural beliefs, popular amongst the broader population as well as political and military figures, means that struggles in Cambodia, like Thailand (Cohen 2012), must attend to the harnessing of the supernatural realm as a political device to evoke both eventfulness and magical effect.

Tying into a feminist geopolitical analytic that emphasises the body and intimacy as sites of resistance to a wider politic (Bosco 2006; Fluri 2009; Hyndman 2001; Smith 2012), a core element of BKL women's strategy to communicate the injustice of forced eviction has been to embody eviction. This is a strategy to which Kevin Doyle makes direct reference; the staging of injustice through and on the body from the imaginative use of props to their physical and symbolic presence in public space on such a quotidian basis. The 'routine' for BKL women and their audiences was to both demonstrate and inhabit the crisis ordinary as strategy.

In May 2012 the vulnerability of BKL women was used as a tool for change when they stripped to their underwear outside the Cambodian parliament.[29] Undressed, their buttocks bore the words 'bare witness' in both English and Khmer (Narim and Zsombor 2012, np). As critical sites of resistance, 'bodies represent humanity in its rawest form. Our bodies are precarious; prone to bruises, breaks, and scars; and reliant on others' (Eileraas 2014, p. 40). Resonating with this, undressing had an analogous potency by revealing, quite literally, stripped rights and the heightened exposure to corporeal vulnerability on the street which contrasted so starkly with the comfort and privilege of officials cool inside their offices. Srei Pov and Heng Mom explained these logics further:

> If we do not have a house to live in, it is akin to a body without clothes on. We are cold and hot with the changes of weather without houses … we wanted to show the differences between them and us, while they are sitting in a comfortable and cooling room. We do not vote for them to sit in a relaxed room without finding resolutions for citizens (Srei Pov).

> Stripping our clothes means there is nothing left which is valuable. Before, we had our houses; we had our private places and rooms, especially for our daughters. Currently, we are faced with eviction and demolition; losing our houses means nothing remains and our values are gone. What's more, stripping the clothes is a message to the government that our villagers are no longer shy. Instead, the government should feel ashamed of their acts (Heng Mom).

The taboo nature of women's semi-naked presence in public space worked to display women's commitment to defending their rights at any cost. Their ruptures to norms tied to Khmer femininity forced spectators to question the motivations for them doing so. As Srei Pov and Heng Mom indicate, their insubordinate and demonstrating bodies warn the Cambodian government that BKL women will transgress traditional gender ideals, including modesty, timidity, and shyness, to make their voices heard and exert agency over their lives.[30] Stripping disrupts the normality of socialising norms for a deliberate pause (Dabashi 2012); the (semi-) naked the body exposes violence by channeling 'bare life' (Agamben 1998). It enacts 'the demand to end precarity by exposing these bodies' vulnerability to failing infrastructure conditions' (Butler, Gambetti, and Sabsay 2016, p. 8) and the gender-based violence of forced eviction.

BKL women simultaneously embodied the personal and political injustices of forced eviction felt through and on their (im)moral bodies to bait the Cambodian government, who they held partly responsible for their suffering (Brickell 2014a). Gender ideals are not always disempowering but rather can have insurgent power, which women harness both to legitimise their actions but also to shame authorities of their very need to mobilise them. Heng Mom, for example, felt their tactic of stripping demonstrated the denial of privacy that was affecting their daughters. The insinuation being made by Heng Mom is that forced eviction takes away more than just the physicality of home; it corrodes the moral infrastructure of the home in which socialisation and societal values are fostered with the next generation.

Aside from stripping, the community accessorised the injustice they were living through. While in Chapter 4 I discussed Chankrisna's wearing of his watch, BKL women took to demonstrating their suffering through hand-made hats balanced on their heads (as also noted in Kevin Doyle's interview). In October 2012 and 2015, BKL women used World Habitat day to make and wear cardboard houses to symbolize the right to have a roof over one's head and women's intimate connection to the homes they were fighting for.[31] In their 2012 demonstration, these houses were used to carry the message 'have home, have life' and to show their common fates.[32] Srei Leap explained more in this regard:

> It is the literal meaning that our houses are our lives. Since they are evicting us, where can we live? We struggle to stay here just to have our houses and lands back. This is the entire meaning we would like to convey to them.

The boxes later worn by the women in 2015 had two windows (or eyes) and one door (the mouth) cut into them to represent a face. The utility of the female body as a living/dying tableau to bring injustice to the forefront of public consciousness was pursued through these annual performances. Gill Drori (2005, p. 176) writes that such UN-dedicated days have garnered wide acceptance both by transnational actors 'as well as by organisations whose aim it is to contest the current hegemony' to the extent that they are 'an arena for cultural praxis: it is where both cultural production and cultural contestation are expressed'. Consistent with this interpretation, the days were used by BKL women as discrete time-spaces to reiterate their claims to home through creative and affective means. Such days are therefore 'punctual' time-spaces in which crisis ordinaries of violence that women encounter in the extra-domestic home are creatively told and packaged on the street within the legitimised wrapper of international days observed by the United Nations. They are notionally 'invited' days of presence that domestic violence victims and other causes have long used (as illustrated earlier), and that BKL women had mobilised in their own repertoire of action. Akin to other eviction threatened communities worldwide then, BKL had a strategy of pursuing recognised channels of redress in combination with dissent forwarded through a micro-politics of improvisation (McFarlane 2011; Vasudevan 2015).

The creative interventions staged on these days are ones that both displayed and deepened women's 'performative labour' (Kanngieser 2013) in their activism against forced eviction. BKL women's performative tactics were mobilised more habitually than the thematic UN days however. The frequency and repetition of these performances over time speaks to the crisis ordinary mode of dwelling in perpetual defence. This included playing on the associational value of Khmer women to peace and non-violence through the repeated conscription of the lotus flower at their regular protests (Figure 5.3).

Figure 5.3   BKL women's lotus-wielding activism, 2012. Source: Courtesy of Erika Pineros.

Phala, the husband of a BKL activist, noted that:

> I observe that women are strong and active in leading the protests. I can imagine as a man seeing a woman holding a lotus flower, crying and describing the difficulties in her life, that the police cannot also emotionally bear to hit women. Women also notice this factor, hence why they decide to stay on the front line. In the meantime, men in the community agree with women, so we all support women. We believe in women's leadership, although it does not mean men are weak. Rather, it means we respect women.

Heng Mom also explained her reading of the lotus motif:

> Concerning the lotus, we would like to say that we all are Cambodians. We follow Khmer ethics, cultures, traditions, and Buddhism. Furthermore, we as women follow the non-violent protest principle. We have no weapons but lotuses. We pray and beg for sympathy from those governmental officers … we hold the lotuses for we would like to express that we are brave enough to solve the issue and dig for solutions. Also, the lotus flower is for worship. Thus, when we confront with police force, we hold up the lotus in front of them; it seems that they are also aware of what we mean. It is the chance we have to explain to those police to learn about their own religion and righteous acts.

Phala's interview excerpt brings to the fore the perceived affective power of women's bodily performances in public space as bearers of peace and pain. The lotus flower was literally and metaphorically held as the motif of women's peaceful protest against the injurious necrocapitalism and accumulation by dispossession of the neoliberal Cambodian economy. As Heng Mom goes on to explain, the lotus was used as a floral compass to morally guide and question officials who attack them at protests. A sacred Buddhist emblem of purity, growth, and transformation, it is waved to legitimate women's public presence and to act as a reminder of shared identities across the barricade line (Figure 5.4). Therefore,

Figure 5.4   Lotus flowers at the barricade line, 2012. Source: Courtesy of Erika Pineros.

while in more isolated instances they sought to highlight their physical vulnerability through indecency not normally tolerated, the lotus became a common visual feature of BKL women's protests to mark their virtuous standing. Just as women have used flowers in other protests around the world (see Betlemidze 2015 on garlands of flowers in FEMEN protests in Ukraine), BKL activists drew on this culturally resonant flower as a means of 'positive' dissent to show ways of being in the world possible apart from violence.

While the examples I have provided of BKL women's creative protest have spoken to the body as a physical frame to prop hats and flowers (and in Ho Lida's case earlier a scarf), it would be remiss not to mention their use of singing to further communicate their predicament affectively. As the media remarks, 'They've wept, they've yelled and they've prayed as they watched excavators tear down their homes, so perhaps it's understandable that the residents of Boeung Kak lake are now turning to song' (Titthara 2011, np). Eight songs were composed by Tep Vanny to honour eight houses that were destroyed in September 2011.[33] Titles included 'Money Destroys the Future', '12.44 Hectares', 'My Fortune and Suffering', and 'Sand Inundates Our Homes'. In 'Development Separates Families', the women sing:

> Tears fall in the middle of the night; my heart is full of pain all the time. With my children we were so warm together but Child, you're still little. Dear, you don't have good fortune.
>
> Mom goes to protest, the children cry and sleep on the ground. Mom carries you in her arms, and caresses you with her tears. Mom is feeling so stifled in mind because we need a home. The land and home is the life of mother for good future. Now she's left the children alone, and nobody takes care of them.
>
> Oh, the development separates us. When mom sees her children running to follow her, she feels so bad. Where is the right of the mother? Please the powerful and rich feel compassion for our home.

The song above was used to vocalize BKL women's position as mothers and to shame the Cambodian government and the men pushing back at them with riot shields. The lyrics speak to the warmth and comfort denied to mothers and their children; their tears and divided hearts. Anna Feigenbaum (2010, p. 367) writes similarly of songs created and sung at Greenham Common Women's Peace Camp in England (1981–2000); that they confront authority by rousing 'deep emotional responses in both singers and listeners [which] generates an affective resonance or reverberation that moves along with the vibrations of the sound itself'. 'Rights talk', or rights singing, in both instances had become a powerful language of advocacy and an effective means to push for change.

BKL also refined an intimate–global strategy attuned to the sites, scales, and temporalities of political power circulating within and beyond Cambodia to make themselves seen as well as heard. In Chapter 4 I described a rooftop near the airport emblazoned with the portrait of then US President Barack Obama and the painted phrase 'SOS' in advance of the 21st ASEAN Summit and 7th East Asia Summit in October 2012. BKL also found bold ways to communicate their domestic emergency and urge US intervention in this opportune moment. On the sand-filled lake facing the Ministry of Councils where the ASEAN Summit was about to take place, balloons were released by the women into the sky at the time they believed Air Force One would land in Phnom Penh with President Obama and Secretary of State Hillary Clinton on board (Figure 5.5). Eviction-threatened communities from across Phnom Penh came together to hold huge plastic banners with the faces of the two leaders and the American flag. Children wore bandanas saying 'WELCOME' and community members lit candles on an enormous SOS sign displayed on the sand that had swallowed their homes. At this moment, the extra-domestic home embodied a vertical geopolitics of home that looked to the skies for action.

Figure 5.5   SOS protest on the former BKL as US President Obama lands to attend ASEAN meetings, 2012. Source: Courtesy of Thomas Cristofoletti.

It is uncertain if these SOS calls were glimpsed by Obama and Clinton as they flew in. What is certain, however, is that BKL women's activism had made their mark on Clinton in particular. Such is the recognition of their determination that in April 2013, Tep Vanny was invited to travel to Washington DC to receive a Vital Voices Global Leadership Award for her work as a human rights defender and spokesperson for BKL (Figure 5.6).[34] Hillary Clinton and then Vice President Joe Biden took to the stage to advocate for the improvement of women's lives around the world. Praise was given for how 'BKL women have tried to rewrite the script affecting communities from across the Global South – developers seize a valuable land, throw the existing community out, and after protests ebb away, a new development arise[s]' (Vital Voices 2013, np). Hillary Clinton told the auditorium of Tep Vanny:

Figure 5.6   Tep Vanny with Hillary Clinton, Vital Voices Awards, Washington DC, 2013. Source: Courtesy of Vital Voices Global Partnership.

I want to thank Vital Voices for honouring a crusader for land rights in Cambodia, a young woman who I had to call several times as Secretary of State to get out of prison because she stood up for a fundamental right that people everywhere should have. Title to their homes, property rights that give them the same stake in the future that everyone deserves to have.[35]

Tep Vanny – depicted in *The Cambodia Daily* newspaper (Mech and Woods 2013, np) as a 'globe-trotting advocate' – shared the stage with the world leaders,

jumping scale from her home under threat in Phnom Penh to the awards ceremony (see Figure 5.6 for their red carpet photo). Staring into the spotlight on
her, she gave her short acceptance speech, 'Our government is like the Khmer
Rouge, but they don't kill with weapons, they kill with corruption. This award is
for all women in Cambodia.' Once again, the imprint of genocide in international
consciousness of Cambodia was being mobilised to draw analogous potency with
the violences of today.

In numerous ways, then, domestic traumas created through forced eviction
have been politicised across multiple scales and times and have challenged the
(in)action of different geopolitical stakeholders. Discrete time-spaces of
global connectivity – from World Habitat Day, to the visit of elite leaders, to
prize ceremonies – offered reminders of BKL women's world presence and
their rights repudiated by a government that only showed a growing intent to
silence them. BKL women's multi-scalar mobilisations and their persistent
creativity over many years had considerable impact. As Kevin Doyle observed
in his personal correspondence with me, 'defiance in a public space. Nothing
a regime fears more than people who have lost their fears.' With this, a governmental and marital backlash followed, which is the focus of the next and final
empirical chapter.

Conclusion

On initial consideration, DV law and forced eviction activism are both refusals for the
intimate wars they embody to be disposed of, or confined to, the domestic. They suggest 'collective thereness' in the wider world (Butler and Anthanasiou 2013, p. 197)
by rejecting public/private and active/passive binaries (Miraftab 2004; Prokhovnik
1998; Yuval-Davis 1997) that have typically dominated influential work on typologies
of citizenship (e.g. Turner 1990). Both DV law and forced eviction activism are events
that exceed their context in time and space. They are both embroiled in a vertical
geopolitics of resistance and citizenship that includes global organisations like the
UN and World Bank and superpowers such as China and the United States in their
storybooks. The research presented in this chapter has, thus far, critically examined
the extra-domestic home and has come to four main conclusions.

First, I have shown the immutability of the distinctions made between invited
and invented spaces of intervention. DV law in Cambodia, I have qualified, sits
ambiguously as an invited terrain of action. Viewed as an unwelcome 'outside'
influence by male parliamentarians who believed it to be detrimental rather than
ameliorative of societal interests, its resistive potential and punitive potency was
truncated on the long journey to ratification. In this sense, the terms and conditions of the invitation matter; the aporetic structure of hospitality meaning that
the 'welcome is never completely free of conditions' (Diprose 2009, pp. 69–70).
More than a decade on from its ratification, the inadequate institutional support

and backing needed to properly implement and enforce the rights discharged in 2005 gives weight to the contention that DV law is a decorative form of 'window dressing' – entertained for strategic purposes to demonstrate compliance to a United Nations rights regime it is officially signatory of yet unamenable to in practice. Combined with the stain of cultural antagonism that DV law – and even rule of law – carries, its existence has not stood in opposition to a bio-necropolitics of abandonment and violence. Rather, the research suggests that DV law, at present, has been viewed and experienced as a tokenistic and incapacitated mechanism of redress. As Chapter 6 goes on to elaborate, the intimate war of domestic violence is being ignored and/or fueled by failures of DV law identified in this chapter that have harmful or even fatal consequences for victims.

Turning to forced eviction, I also showed how women's unsanctioned activism and oppositional consciousness in BKL is highly contingent and ambivalent. According to Faranak Miraftab (2009, p. 35), 'insurgent movements do not constrain themselves to the spaces for citizen participation sanctioned by the authorities (invited spaces); they invent new spaces or re-appropriate old ones where they can invoke their citizenship rights to further their counter-hegemonic interests'. BKL mobilised dominant norms and invited tropes of Cambodian womanhood that were normative in reference but subversive in performance. Just as the DV law research showed a strong rhetoric towards the inalienability of Khmer culture and customs, which were crystallized into the law as a condition of its being, BKL women appropriated and retooled ideals of Khmer womanhood as domestic guardians and mothers to assert their citizenship rights.

Second, in this chapter I have revealed how gender norms were promoted in both cases as protective infrastructures as much as subordinating ones in women's lives. Therefore, while in the case of forced eviction, I explored how BKL women contested the actions of the Cambodian government day after day, year after year, the chapter was equally careful not to overstate their counter-hegemony. BKL women thus occupy a liminal position; as unruly subjects who protested the dependency of necrocapitalism on their submission, at the same time as obedient female citizens taking responsibility for their children and family through the defence of their homes. Much like for the Patronas in Mexico, these outcomes are not mutually exclusive; rather their combination acts as means to 'weaponize domesticity' (Bruzzone 2017, p. 249) amidst the intimate wars they face.

Given the range of cultural, economic, and physical risks that domestic violence victims encounter in any legal claims against intimate partners, when pursued they are also engaging in an assertion of oppositional consciousness. This is because this decision goes against the norm of non-reporting and treating male violence with impunity. Decisions taken equally to end their marriages are resistive in the sense that these are 'acts' of citizenship (Isin 2008; Mukhopadhy 2007) that are enlivened engagements with the right to divorce, and ones that the Cambodian body politic would rather remain abstract. A woman's move into 'widowhood' could even be classed as an insurgent one given the painting of marital breakdown and the perception of DV law as a dangerous space of citizenship (see Holston 2009). This is because should these rights be pressed, and gender equality forged, there was a risk of social, cultural, and economic wounding to multiple bodies of the victim, home, and nation. Albeit unable to transform the hierarchical and patriarchal foundations of society, in such instances of separation and divorce, women have deliberately gone off script.

What marks out BKL women's endeavours from decisions taken by domestic violence victims, however, is their committed level of creativity and deliberate intent to rewrite the script and recreate the scene of their dispossession on the national and international stage. Domestic violence survivors and many of the institutional stakeholders I interviewed try to manage or minimise the regularity and severity of violence encountered within the 'effectiveness' (or ineffectiveness) of the existing order. Forced eviction activists in BKL, by contrast, show a commitment to 'disruption' viewed necessary for meaningful change (Buire and Staeheli 2017). In other words, while women's help-seeking in the context of domestic violence could be understood as small acts constitutive of a 'quiet politics' of activism given the risks involved (Pain 2014), the 'loud' action taken on the streets of Phnom Penh by BKL women is intended to evoke an oppositional consciousness that encourages other communities to call time on necrocapitalist plots to evict them. This is not to say, however, that their survival-work should be viewed only through an antagonistic and demonstrative lens. An expanded academic account of activism also includes quotidian acts of connectivity, creativity, and even cultural compliance (Pottinger 2017), which were also on ample display in the chapter.

Third, related to this, actions taken to address domestic violence and forced eviction speak to the complexity and slippage of citizenship as status (formal membership in a particular political community) and citizenship as practice (active participation). Political economy dynamics, which I encapsulated under the heading of precarious economics, mean that many women who have experienced domestic violence are deterred from pursuing their legal rights given the multiple risks of doing so. DV law does not help negate the problem of asserting rights only to experience 'social death' as a consequence of marital breakdown that precedes or ensures their use. Moreover, it could be argued that DV law has expanded and legitimised the administrative apparatuses of state violence to punish women in these socio-economic terms, for trying to make a change in their lives. As such, the precarity laced through women's daily life 'renders transformation contingent and difficult' (Ettlinger 2007, p. 321). On pragmatic grounds, the majority of women's experiences of lived citizenship as domestic violence victims are characterised by their everyday needs and labouring for survival rather than a resistive turn to DV law as a (futile) carrier of hope.

In the *Progress of the World's Women* (UNWOMEN 2011) report titled 'In Pursuit of Justice' it is cautioned that 'for most of the world's women, the laws that exist on paper do not translate to equality and justice'. Going from paper to practice means, UNWOMEN argue, ensuring that a 'functioning justice chain' of processes and institutions exists to support women's rights-claiming. Narratives on the impotency of law in Cambodia more generally suggest that this justice chain will remain broken long into the future and the actual and perceived dangers of seeking legal redress will remain stark unless structural political change occurs (see Chapters 6 and 7 for evidence of the diminishing likelihood of this). Law in the Cambodian context could be considered a form of 'cruel optimism' (Berlant 2011) and perhaps even cruelty. In other words, while the prospect of DV law instilled a discursive sense of possibility, it is marked in actuality by its impossibility given indirect failures to protect women from the violences it originally or performatively sought to outlaw.

While in the case of DV law I have been rather cynical in my conclusions here, I am not blind to the positive impacts of legal commitments however empty observance is. By focusing on BKL women's activism during the 2008–2012 period it became evident, for example, that international human rights in a weak regime have the capacity to empower 'nonstate advocates with the tools to pressure government towards compliance' (Hafner-Burton and Tsutsui 2005, p. 1378). In this sense, resistance calls on multiple scales of activation and agitation from local to global which women harnessed for their own strategic ends. BKL women's intent and actions in their cogent repudiation of 'development' and holding to account of the World Bank extols perhaps the 'emancipatory potential of feminism', which depends on 'reconnecting struggles against personalized subjection to the critique of a capitalist system' (Fraser 2009, p. 388).[36] Fundamentally, however, both domestic violence and forced eviction victims are systematically encountering a structural violence of denial; not only from the right to dwell, but also from citizenship and legal entitlements that they are de facto excluded from claiming without the threat of other administrative and intimate violences being wielded in response.

Fourth, and finally, the chapter has highlighted the spatialities of gendered struggle within and against myriad types of violence in women's lives. Despite the claims made by elite government officials that DV law has revisioned violence against women in the home as a societal issue, the law ratified does not reflect this but rather reconfirms the necessity to keep 'fire in the house' in order to ensure the fallacy of familial harmony. BKL women's activism, meanwhile, is a high order irritant because it brings state-enforced familial *dis*harmony into the public domain that has a political and economic cost. As Chapter 4 established, if women fail to control 'fire in the house' and/or it breaks up, their homes and bodies can be considered truncated. Forced eviction is a 'fire' that BKL women are trying to put out within their threatened homes, but its fuel resides outside and for this reason requires they also step outside to fight it. In the next chapter I turn to attempts to maintain power and truncate the stories of BKL women who have ventured and laboured beyond the boundaries of home to defend the survival of their families in the crisis ordinary that has come to define contemporary Cambodia.

Endnotes

1 According to the World Bank (2012, p. 336), 'The Cambodia Committee of Women, a coalition of 32 nongovernmental organizations (NGOs), persistently lobbied the government and the Ministry of Women's Affairs to secure the legislation's passage.'

2 The formulation of these new laws should, according to UNWOMEN (2012, p. 1), adopt 'a comprehensive legislative approach, encompassing not only the criminalization of all forms of violence against women and the effective prosecution and punishment of perpetrators, but also the prevention of violence, and the empowerment, support and protection of survivors. It recommends that legislation explicitly recognize violence against women as a form of gender-based discrimination and a violation of women's human rights.' The handbook makes clear that this approach and 'model framework' it provides is essential if such legal interventions are to be effective. The Beijing Platform for Action, adopted at the Fourth World Conference on Women in Beijing in 1995, also called on governments to enact and reinforce penal sanctions in domestic legislation. This call was reiterated during the five-year review of the Beijing Platform for Action in 2000.

3 Moreover, as Peg Levine (2010, p. 32) concurs, in Cambodia 'love within marriage is often seen as a bonus, while family harmony is the cake of tradition'.

4 FUNCIPEC is a royalist political party in Cambodia which was founded by Norodom Sihanouk in 1981.

5 Dr Kek Galabru (Pung Chhiv Kek) is President of the Cambodia League for the Promotion and Defence of Human Rights (LICADHO). She played a significant role in bringing DV law to fruition with the community of NGOs aforementioned.

6 http://primeministerhunsen.blogspot.co.uk/2011/03/address-at-celebration-of-100th.html.

7 This is significant given that in Article 9 of DV law it is stated that 'The nearest authorities in charge have the duty to urgently intervene in case domestic violence occurs or is likely to occur in order to prevent and protect the victims. During the intervention, the authorities in charge shall make a clear record about the incident and then report it immediately to the prosecutors in charge.'

8 See www.onebillionrising.org.

9 See https://www.onebillionrising.org/9958/rising-statement-cambodian-v-activists.

10 See https://witness.theguardian.com/user/Sangiec.

11 A recent LICADHO (2017) report detailed 392 cases of domestic violence (237 now closed) that the NGO investigated between the beginning of 2014 and the end of 2016. The report noted relatedly that: 'In almost all cases investigated by LICADHO, the couples involved are poor or very poor and their economic survival depends on their having at least two adults to provide financial and other resources as well as child care. There is no state provided social security system in Cambodia and other than a few NGOs providing limited safe shelter to victims of domestic violence, there is no provision to assist women leaving a violent situation. It is therefore practically very difficult for them to do so' (p. 11).

12 Indeed, Transparency International's 2017 Corruptions Perception Index released in 2018 marks Cambodia out as the most corrupt in Southeast Asia, and in 161st position out of 180 countries worldwide (with a score of 21 out of 100 points). As Transparency International (2017) outlined on their website, 'The lower-ranked countries in our index

are plagued by untrustworthy and badly functioning public institutions like the police and judiciary. Even where anti-corruption laws are on the books, in practice they're often skirted or ignored. People frequently face situations of bribery and extortion, rely on basic services that have been undermined by the misappropriation of funds, and confront official indifference when seeking redress from authorities that are on the take.'

13 1000 riels equates to approximately US$0.25.

14 Examples of such political corruption also extended to the withholding of family discount cards and school scholarships in the DV law study.

15 At the time of writing, the Cambodian People's Party (CPP) is the ruling party. In 2012 the Sam Rainsy party merged with the Human Rights Party to become the Cambodian National Rescue Party (CNRP) – the CPP's main opposition until it was dissolved in 2017.

16 They support her study at beauty school (US$300/month) and monthly outgoings (US$60).

17 Research by LICADHO (2017, p. 2) has found that in its closed cases, there were three main categories of survivor: 101 (43%) ended with the victim deciding to drop the case and to return to live with her violent husband; 69 (29%) ended with the couple separating or divorcing but with no criminal case being brought against the perpetrator; and 53 (22%) ended with a criminal trial. Another 14 (6%) of cases ended in other ways, for example the perpetrator committed suicide (LICADHO 2017, p. 2). Of the 101 victims who decided to drop the case and return to their husbands, 34 (34%) made the decision without the intervention of an official, 19 (19%) reconciled after a negotiation facilitated by the village or commune authorities, 30 (30%) reconciled after a negotiation facilitated by the police, and 20 (20%) were reconciled by a civil court judge during their divorce proceedings (LICADHO 2017, p. 4).

18 See E. Wayne Ross and Kevin Vinson (2013) for work on pedagogies of resistance and the teaching of the 'dangerous citizen'.

19 While the *Chbap* were written between the fourteenth and nineteenth centuries and passed down from generation to generation, one version of the *Chbap Srei* was composed in the mid-nineteenth century by King Anh Doung and modelled on a lecture by the Buddha (see Derks 2008). The most recent version, written by former monk Min Mai, also became the taught version memorized in schools (Derks 2008). See Jacobsen (2008, p. 119) for a discussion of the contested authorship of the *Chbap Srei*. See also Ledgerwood (1990), Ovesen, Trankell, and Öjendal (1996), Ayres (2000), and Luco (2002).

20 See Alexandra Gattrell (2010) who writes about the experiences of disabled Cambodians and whose participant captured his sense of entrapment through the use of this proverb.

21 In the article, Tep Vanny's (then) husband, Ou Kongchea, is photographed with a feather duster taking care of domestic duties in their home.

22 As Chapter 6 goes on to explore, however, such accommodations of women's public protest did not always last given the long duration of their activism, and some marriages have since ended.

23 This possibility is also noted in Jonathan Darling's (2017, p. 189) work on the precarity of asylum seekers in which he explains that a 'concern with urban informality enables a valorization of incremental and often highly tactical practices that can constitute "minor" political acts'.

24 In June 2011 around 200 BKL residents also gathered outside EU offices as a petition was submitted by community representatives.

25 These road blockages, usually of Monivong Boulevard, continued to take place with multiple examples reported in the press during 2013.

26 Police brutality and (the threat of) incarceration also grew over this period and is covered in detail by Chapter 6, which focuses on the consequences of their actions.

27 See comments by Erik Davis in this media piece on differences between Brahminist and Buddhist practices used by Boeung Kak women. Further news coverage of their cursing rituals is available (Narim 2013; Narim and Crothers 2014).

28 The Preah Vihear temple is an eleventh-century world heritage site located on the Cambodia–Thai border. Its sovereignty remains disputed despite a 1962 ruling at the International Court of Justice (that both countries accepted at the time), which saw Cambodia awarded the Hindu temple.

29 See the film documentary *Every Bird Needs a House* (Trintignant-Corneau and Chansou 2013) for video evidence of this.

30 Other taboo actions were taken. On 10 February 2011 houses in village 22 of BKL were threatened with eviction. BKL women threatened to throw their period blood at the police if they attempted the eviction (SaveBoeungKak 2011).

31 Hundreds of residents from across evicted-affected communities in Phnom Penh also met in 2010 to mark World Habitat Day.

32 At this event the women joined around 400 representatives of 40 communities affected by eviction (including Dey Krahorm).

33 For further information, see: http://licadhocanada.com/media-room/songs.

34 See my newspaper articles for further commentary (Brickell 2013a, 2013b).

35 Audio transcribed from *A Cambodian Spring* (Kelly 2018) film.

36 However, the lending freeze that began in August 2011 has reversed after five years. In May 2016, the World Bank approved US$130 million worth of new loans to Cambodia (Paviour 2016). This was met by anger from many national NGOs, international groups, and Tep Vanny herself.

Chapter Six
Intimate Wounds of Law and Lawfare

Introduction

In this chapter I explore the amplification of violence in the lives of Cambodian women through law and lawfare and the prisms of inequality, poverty, and patriarchy that they refract. '"Law" draws lines, constructs insides and outsides, assigns legal meanings to lines, and attaches legal consequences to crossing them' (Delaney 2015, p. 99). This is an observation that is especially pertinent in Cambodia given refusals made by some victims for the intimate wars of domestic violence and forced eviction to be disposed of, or confined to, the private. The chapter charts some of the outcomes of these refusals and the mediating influence of law and lawfare.

In doing so, the chapter brings into dialogue discussions of lawfare with that of intimate war as a core concept and lived experience fused through the pages of *Home SOS*. While touched upon in Chapters 4 and 5, I take a closer look at the weaponisation of law against women's homes and bodies to extend my analysis of its perpetration and everyday politics. Cambodia has both a 'thin rule of law' as part of its institutional culture as well as 'rule by law' in which the law is enforced 'by the executive judiciary, or various arms of the state without regard for, or at least expense of, justice' (McCarthy and Un 2017, p. 103). As the Special Rapporteur on the situation of Human Rights confirms, 'Cambodia can be characterised as increasingly being ruled by law' (Rhona Smith cited in Coffey 2018, p. 206). I demonstrate, as a result, how the Cambodian government uses

Home SOS: Gender, Violence, and Survival in Crisis Ordinary Cambodia, First Edition. Katherine Brickell.
© 2020 Royal Geographical Society (with the Institute of British Geographers). Published 2020 by John Wiley & Sons Ltd.

a strategic combination of law (its omission) and targeted lawfare (its commission) as twin strategies to consolidate power and suppress women's dissent against these intimate wars. In their discussion of gender-based violence, including femicides, in Honduras, Cecilia Menjívar and Shannon Drysdale Walsh (2017, p. 223) set out the conceptual parameters of such omissions and commissions in more detail:

> Even if the government is not killing women directly, acts of commission and omission create conditions that promote impunity and increase risks of victimization by normalizing the targeting of women for violence, at home and in the streets. Though acts of omission may not directly involve the state in the killings, inaction can also lead to such killings.

Noting their indivisibility, the scholars go on to argue that both omission and commission 'have roots in the same social context that normalizes and sustains violence' and that this 'context also shapes the lens through which state actors assess information, justify acts, and implement laws' (Menjívar and Walsh 2017, p. 222). Taking these nuances and qualifications further, acts of commission (e.g. direct forms of violence) and those of omission (e.g. structural violence) fold in on each other (Anglin 1998; Confortini 2006; Fryberg and Eason 2017; Tyner 2012; Tyner and Rice 2016). 'When we recast structural violence within the context of letting die', James Tyner and Stian Rice (2016, p. 48) explain, 'we readily see that many individuals, such as politicians and corporate managers ... (1) are aware of harmful policies and practices that might disallow life; (2) have the opportunity to stop or remedy these policies and practices; and (3) have the financial – or political – ability to prevent harm'. Under these preconditions of awareness, opportunity, and ability, a failure to act can be considered intentional (Green 1980; Tyner and Rice 2016). Structural violence is then a commissioned omission. Commissions and omissions are both agentic instantiations of intimate warmongering against women whose potential to survive, to live, is diminished. The ontological slippage between 'killing' and 'letting die', which I briefly considered in prior chapters, is therefore elaborated upon in the present one through the cross-fertilization of both omissions and commissions into my analysis of the outcomes of the SOS calls (Chapter 4) and responses (Chapter 5). The chapter thereby responds to the call made in geography for attention to be given to how 'in-action generates its own harms' (Anderson et al. 2019, p. 5).

By studying domestic violence and forced eviction together, I illustrate how women's management of these omissions and commissions are further part of the crisis ordinary that they inhabit. Synergies can be made back to previous chapters and Achille Mbembé's (2003, p. 21, emphasis in original) seminal work on being 'kept alive but in a *state of injury*, in a phantom-like world of horrors and intense cruelty and profanity'. Active inaction by the state becomes a means of letting die given limited opportunities or freedoms to improve one's situation (Davies, Isakjee, and Dhebi 2017; Tyner 2016b). Under these conditions, 'excluded groups may not be actively killed but are instead allowed to suffer the brutal indignity of harmful spatial environments' (Davies et al. 2017, p. 8). In a connected sense, I go on to show how women are being left to die, or else manage the crisis ordinary of their wounded homes and bodies, which can never be fully healed without structural and political change. This again evokes the suggestion made by Silvia Federici (2018b, p. 53) that 'micropolitics mimic and merge with institutional macropolitics' to preclude women's freedom from violence.

In this final empirical chapter, I once again divide my analysis between the domestic violence and forced eviction case studies. In the first part, I explore how the abject failures of laws governing domestic violence have resigned help-seeking women to their normative place at home and in the second, I showcase the waging of law against BKL women by state adversaries with a strong focus on events since 2012. Across them I demonstrate how lawfare – as a hybrid of legal omission and commission wielded against its enemies – is being used to delegitimise and harm women who challenge governmental agendas and their monopolies of power in Cambodian society today. I also show how non-violent legal advocacy and grassroots activism in both domestic violence and forced eviction cases is met with the violence of lawfare.

What will become noticeable is that the weight of the chapter tilts towards the latter case study, of forced eviction and activism rallied in response. Rather than artificially or aesthetically adhere to the balance that I achieved in Chapters 4 and 5, I embrace this asymmetry on three grounds. First, domestic violence (law) remains still locked into the status quo, as demonstrated by the attritional violence and survival-work that women are compelled to still undertake. It remains an all-too-ordinary violence that feels too routine, too normal, to be subject to transformative change. While the passing of domestic violence law (DV law) in Cambodia was a long and politicised process that kept the spectre of domestic violence in view for several years, its fading eventfulness in the public realm does not reflect the ongoing crisis ordinary perpetuated in the extra-domestic. By comparison, the BKL case of forced eviction activism is arguably more extraordinary, and certainly more exceptional, in its break from the status quo and instructive mantra 'Don't hit a stone with an egg'.

A second and inter-related point is that in chronological terms, BKL women's actions have continued to provoke overt and highly charged political reactions that domestic violence advocacy has not. The longevity of its public eventfulness contrasts with the relative uneventfulness of domestic violence (law) in the public domain. The imbalance of the data from the two case studies reflects the fact then that social struggles have different temporalities and variable levels of eventfulness. The analysis I provide on their respective crisis temporalities is only made possible by the longitudinal nature of my research engagements, and reflects the differential paths, durations, and intensities of violence in Cambodia today.

Third and finally, I believe that the emphasis of my analysis on forced eviction strengthens rather than sidelines the significance of domestic violence. This chapter witnesses the collapsing in, rather than separating out, of domestic violence and forced eviction as gender-based violences simultaneously experienced in marriage by BKL activists. As such, and as befitting of the final empirical chapter, I signal the value of writing a cross-issue book that synergises and mobilises a more complex and integrated understanding of intimate geographies of violence than would otherwise be possible.

The Law and Lawfare of Domestic Violence

In what follows I focus on the modus operandi of local reconciliation, which continues to be venerated on moral grounds by local authorities and in the context of a compromised judicial legal system that women find themselves pitted against. Despite Article 27 in the DV law, which states that reconciliation is not meant to put 'pressures on the party who refuses to go along with each other', I show how too often the imperative placed on harmony, consensus, and the collective means that heavily injured victims are still pressured into reconciliation with the same husband, often on a repeated basis. Taking this argument one step further, the legitimacy that DV law gives to reconciliation obscures the power relations which permit that imposition to be successful.

Local reconciliation, I continue relatedly, represents a compensatory justice mechanism that conceals injustice and domination. It is a diversionary process that distracts from, rather than offers solutions to, the compromised and seemingly intractable political conditions it is deployed within. Local reconciliation represents a falsely purported and potentially harmful move from violence to peace for its victims. It does little to bring about structural change, egalitarian social relations, and social justice – prerequisites for 'positive peace' (Galtung 1969).[1] Rather, it is an example of 'negative peace' (Galtung 1969) – a peacemaking enterprise that is uncertain and commonly unfinished. The crisis ordinary that many Cambodian women are navigating is this unconcluded journey of violence woven throughout their lives. The gaping wounds of domestic violence that remain, be these physical and/or emotional, are those of the crisis ordinary that women tend to. With this in mind and moving forward, I focus, first, on the philosophy and performance of local reconciliation and, second, on its (potentially) fatal implications.

Reconciled to Violence

Samroh samruol – meaning to smooth over and seek harmony – holds a continued significance in contemporary Cambodia as a reconciliatory approach that promotes the administering of justice where conflicts arise. The commune office is the site of this rehearsed consensus and the pretence of harmony in which familial expectations about gender norms usually prevail (Figure 6.1). That Article 9 of the DV law states that 'the nearest authorities in charge have the duty to urgently intervene in case domestic violence occurs or is likely to occur in order to prevent and protect the victims' has a scalar pertinence in this regard. In local reconciliation, a lot of effort is invested in avoiding open, public disputes that would damage pride and bring shame, not only for the family, but also for the community. '*Somroh somruel* mediation and the value it places on balancing the needs of the individual with those of the collective' (Ramage et al. 2008, p. 2) means that pressure is on none of the parties to 'lose face'. As a no-fault system, women also risk being undeservedly blamed with their needs subordinated to the privileging of others.

Figure 6.1   A typical commune office where local reconciliation takes place, 2012. Source: Katherine Brickell.

Under these circumstances, it could be argued that local reconciliation aggravates the potential for symbolic and administrative violence through the de facto hushing of women's credible legal rights in favour of 'resolution'. As Pierre Bourdieu (1987, p. 812) posits, 'Symbolic violence implies the imposition of such principles of division, and more generally of any symbolic representations (languages, conceptualizations, portrayals), on recipients who have little choice about whether to accept or reject them.' For many women there is a durable effect of this suggestion; that domestic violence victims reconcile with their perpetrator. While reconciliation is usually associated with the redemptive power of listening, the offering of remorse and forgiveness (Bregazzi and Jackson 2018), this traditional technique of intimate governance has itself the potential to be violent. By allowing local reconciliation in DV law, it could be argued that the state is complicit in the symbolic approval of patriarchal violence internal to its own reproduction. Such reference to harmony and its putative potential is not only observable in Cambodia, but also elsewhere in the region, including China. Liljia Xie, Eyre, and Barker (2017, p. 8) explain that in China a discourse of social harmony is commonly used to coerce women back to dangerous homes, thereby relying on 'an imagined kind of femininity that could obliterate all violence with self-sacrifice' and reproduce 'the primary designation of women as the guardians of a stable home'. This observation mirrors patterns evident in Cambodia (described in Chapters 4 and 5) where the onus is less on the upholding of legal duties in protecting women but more about propping up moralistic and nationalist discourses of harmony promulgated by the state. Figures from the Ministry of Women's Affairs (Committee on Elimination of Discrimination Against Women (CEDAW) 2018, p. 16) show that in six of the 12 provinces with the forms and systems in place, between 2014 and 2016 only 19 protection orders were issued. This also reveals that the administrative apparatus is still missing in half of Cambodia's 25 provinces for women to be able to file a complaint to provincial/municipal courts asking for issuing a protection order (as per Article 16 of DV law).

On the ground, local reconciliation is ordinarily sought through a meeting orchestrated by a village or commune leader who tries to encourage compromise between parties to reach an agreement marked verbally or by a promissory note. Proponents of restorative justice argue that this 'can be effective in the instrumental sense of reducing the likelihood of reoffending, and effective in the symbolic sense of occasioning strong censure in the individual case' as beyond social tolerance (Hudson 2002, p. 626). The promissory note is aimed, therefore, to inspire fear, which is believed to catalyse the perpetrator to question his violent actions. Interviews with Chankrisna, Ponlok, and Rangsey support this point:

> The legal procedure does not require anyone to sign a note. It is done mostly for the psychological effect. It seems to do the trick in most cases. It is nothing but a scare tactic done by the authority (Chankrisna, 31-year-old male lawyer).

> I think that it [the note] is a good method used by the authority trying to curb the issue of domestic violence. Think about it this way, these men are not very educated in the first place. They will be more prone to fear. The note can be used to subdue their aggressive tendencies and scare them into obeying the law. You may say that it is not the primary goal of this law and you may be correct in your assessment. However, if you consider the number of those men's spouses spared from the beating because of this little note, I think that the good will outshines the bad almost all the time. I understand that this is not the best way because those violent people will not care about this note. Some still break the promise they made earlier (Ponlok, 29-year-old female legal clerk).

> The promissory note details information and promises made by the perpetrator to guarantee the safety of the victim. There is another problem that I witness here as

well though. Even after these perpetrators break their own promises, the authority does nothing to stop them … and the note signed is not really from their [perpetrators] hearts. The authority tends to use the note as a scare tactic (Rangsey, 35-year-old male NGO worker).

Reference to the promissory note as a 'scare tactic' was commonplace parlance in the interviews, including the three I include here. Together they speak to the micropolitics of fear that are attached to domestic violence and its mitigation management and performance (Pain 2014). While reliance on these personal statements is justified in the majority of cases on grounds of moral rectitude, the dysfunction of the Cambodian legal system means that some institutional stakeholders use the promissory note as an unsatisfactory yet compensatory means for managing domestic violence under constrained professional circumstances.[2] They are conjuring acts or 'tricks'. The notes instil a temporary sense of hope for both the victim and authority figure that the violence will cease within the constraining environment they inhabit.[3] At the same time, however, they are both pragmatic enough to realize that the cycle of violence is unlikely to be broken. Not only will the perpetrator go on to commit further acts of violence but the promise of DV law will be undermined by the lack of action taken when this occurs.[4] The symbolic violence that local reconciliation enacts is not only one of omission, by failing to support women's legal rights, but also one of administrative commission, in which the promissory note is directly wielded to 'patch up' a violent marriage. This 'karma approach' to reconciliation (Galtung 2001) pursues 'the idea of cooperating to plug the holes in the boat that we share rather than searching for the one who drilled the first'. In other words, the promissory note is the plug being used to stem the flow of domestic violence for which the couple share responsibility. This approach to local reconciliation rarely succeeds and fails to support women's right to dwell free from violence.[5]

Ultimately noting 'limited progress in the prevention and elimination of violence against women', *samroh samruol* has been criticized by the United Nations Committee on the Elimination of Discrimination against Women (CEDAW 2013, p. 4). In advance of Cambodia's monitoring, the working group had asked for information on measures taken by the government to counter the use of local reconciliation processes in dealing with violence against women. They noted:

> The Committee, while noting the explanations provided by the State party regarding its practice of *disposing* of cases of violence against women through mediation, is concerned that this may discourage women from taking legal action against perpetrators, even when such action is warranted (CEDAW 2013, p. 5, my emphasis).

While a negligent state may 'avert its gaze' (Scheper-Hughes 1992) and fail to implement laws to protect women, the allegation is that there is more active and deliberate disposal of women's rights via local reconciliation. Taking this argument one step further, there is an intentionality that lies behind the generation of impunity by omission. This also means that 'active impunity mechanisms that suspend the applications of the law are not necessary if the law has not, or indeed cannot, recognize either the offender or the harm in question in the first place' (Liu 2015, p. 16). This is a cogent point in respect to Cambodia's DV law and Article 26 in particular:

> For the offences that are the mental/psychological or economic affected violent acts and minor misdemeanors, or petty crimes, reconciliation or mediation can be conducted with the agreement from both parties. The household members can choose any way by requesting parents, relatives, Buddhist monks, elders, village chiefs, and commune councillors to act as the arbitrators to solve the problems in order to preserve the harmony within the household in line with the nation's good custom and tradition in accordance with Article 45 of the Constitution of the Kingdom of Cambodia.

Article 26 sets out that the law does not condone local reconciliation; rather it endorses it in certain circumstances for the benefit of the national collective. While both parties – the victim and the perpetrator – should agree to participate, household members are positioned as their proxy. They are provisioned with the agency to make the decision to participate, seemingly with or without consent from the victim, whose own interests and needs may not align. While women have a right to refuse, 'securing a right to someone requires making the person really capable of choosing that function' Martha Nussbaum (2005, p. 175) moderates. Such concerns have been reiterated by UN Special Rapporteur Rhona Smith, who observes that there has been no evidence that legal provisions have translated into greater protection for women in Cambodia. One of her reports states that:

> While the law on the prevention of domestic violence only permits mediation in cases that are not criminal, the Government acknowledges that mediation is commonly used, resulting in many victims of domestic violence returning to live with an alleged perpetrator who has not been criminally investigated or prosecuted. Mediation and informal mechanisms can result in the retraumatization and degradation of the victim, who may then find herself still living with the perpetrator. Further physical and mental violence is possible (United Nations General Assembly 2016b, p. 8).

In Western case studies, which comprise the vast majority of scholarship, local reconciliation is also discussed as a potentially dangerous proposal for DV treatment (Kohn 2010; Lewis, Dobash, and Cavanagh 2001; Stubbs 2007). A seminal paper by Ruth Lewis et al. (2001) outlines a compelling set of arguments of why community justice is problematic: that an overly positive notion of consensus is adopted; that the intent, beliefs, and reasons for men using violence are ignored; that most DV victims will be scared of their perpetrator(s) and the potential for retaliation; and the power asymmetries between the two remain unopposed.[6] On this basis, the authors advocate for the formal accountable legal system to be used so as to relieve women of the pressure to challenge men's violence.

Yet this is a challenge in any country, but *especially* in Cambodia, where currently an accountable legal system is a vain, even cruel, hope. In their work on rape in the country, for example, Catherine Burns and Kathleen Daly (2014, p. 65) identify 'a clash between human rights rhetoric – that victims should report rapes to the police and perpetrators should be punished – and the realities faced by rape victims and their families – that greater shame will occur by engaging with the criminal justice system, which in any event is corrupt and beyond the economic reach of all but an elite'. In Chapter 5 I also showed how domestic violence victims had experienced revictimising dynamics in the Cambodian legal system that echoed those that they had experienced from their husband, namely silencing, coercive control, and economic violence. If marital breakdown ensues then many women feel a sense of conflicting loyalty, to their children, to norms and ideals, and to their wider community. Yet all of these risks do not negate the symbolically coerced participation and inadequate attention to the victim's safety that I have outlined in this section and that I continue to explore in the next.

Fatal Wounds of Local Reconciliation

> Physical harm tends to serve as a defining point for us, really. The issue becomes criminal and it affects our society greatly. You may ask the wife if it is OK for her husband to physically abuse her. She may say that it is permissible for this one time because he was under the influence of alcohol. It is partly my fault as well. If he sees the mistake and promises to stop his folly, then all that is needed then is a promissory note. We can help accommodate that practice. However, when serious injuries occur, then the reconciliation process is thrown out of the window. The perpetrator must now face the consequence of his actions. There is no exception for that (Oudom, 55-year-old district chief in charge of administering seven communes).

As already indicated, in cases where the victim has experienced 'criminal offences that are characterized as felonies or severe misdemeanours', under Article 17 of the DV law, authorities in charge should not intervene to reconcile or mediate; rather equivalent articles in the country's Penal Code should be used to pursue a criminal conviction.[7] Despite the strong assurances that Oudom gives that local reconciliation is 'thrown out of the window' when physical violence is reported, perpetrators are not usually held accountable in law.

This was the case for Lida who was informally married via a customary ceremony in 2012 to her second husband Nhean, a construction worker. Whilst initially unaware of his violent past, she soon discovered that his previous wife ran away after years of abuse. Although caring at the start of their marriage, his behaviour soon changed. A casualty to marital rape on an almost daily basis and with her clothes all burned by Nhean, she describes her marriage as a 'joke'. She has also been subject to severe physical violence and has been hit variously with a pestle and mortar, hammer, and belt – the weaponry of intimate war.

> The last time he hit me so hard that I cracked my skull … he swung a samurai sword over my head … when I came back from hospital, my intention was to just collect my stuff and go. I told the commune I needed a divorce, but the authority told me to go home first. They wanted us to reconcile … I thought long and hard about it and realised that I didn't want to have too many husbands. It's not a good thing. So, I decided to forgive him and we got back together. Not even a week after that, he did it to me again. He punched me as if he was a boxer … the community won't intervene. I can get beaten to death, and they won't care. It's like I have no rights at all.

Lida's experiences highlight local reconciliation as the ruling mode of dispute resolution.[8] She emphasises the unsolvable nature of her situation dwelling within an environment of impunity in which disinterest and inaction would only come from her imagined death. By extension, Lida believed she was not credited as a life worthy of saving (see Butler 2009). In other words, her killing would be seen as admissible; a 'letting die' prefaced on a mantra of non-interference and privacy that continues to structure Cambodian everyday life and the very survival of domestic violence victims. This is despite claims made to its societal reorientation by government representatives (see Chapter 5).

Attempts to pursue a criminal case against Nhean were incorrectly rebuffed by commune authorities. Not only this but the potential source of disadvantage and decreased status that Lida would face by leaving Nhean encouraged her to follow an aunt's advice to remain 'for the sake of the children'. While Lida has been abandoned by those charged with the legal responsibility to protect her, she has been left with an overhanging sense that *she* had *angered* local officials on account of her repetitive help-seeking. In Chapter 5 I described Tola's similar experiences of being denied help. In the interview, Tola went on to tell me about the chocking to death of her friend and the significance of local reconciliation in this fatal wounding. Having filed multiple complaints prior to the death, but asked for money each time, Tola describes a 'back and forth' situation in which her friend was told repeatedly to reconcile. As Rob Nixon (2011, p. 16) writes briefly in respect to domestic violence, 'it may be life threatening but slow, bloodless, and brutal in ways that are not always immediately fatal'. Incremental failures can be fatal, however, as the eventual murder of Tola's friend demonstrates. As a female NGO leader told me in a conversation we had on the promissory note:

> Honestly speaking, these agreements are meaningless. The mediator thinks that he might be able to reduce the domestic violence. However, if at any point the party causes an injury, it will be an assault case and it is automatically governed by Criminal Law. The case should be brought to the court. As I told you, however, it is such a traditional practice in Cambodia; the case will be unresolved and each time

the violence will be forgiven. Finally, the wife might be killed … I am used to seeing these notes as many as ten times.

DV law has enabled local reconciliation to become a sanctioned means to administer a form of deterrence that is not only futile for many women, but can also be fatal. The promissory note is thus the violent matter of administrative lawfare waged against women on an often-repeated basis and sometimes with deadly consequences. In this sense, the crisis ordinary is the 'unresolved'; the unrelenting until death. Rangsey, who I introduced earlier, also went on to make connections between the propensity for local reconciliation and the murder of women living in protracted circumstances of domestic violence:

> There is one family here that went through a horrendous time, which ultimately led to the death of the wife [Roumjong]. Initially, they engaged in violence between each other. The wife would get drunk after taking the beating. Once, the police arrested the man after a huge altercation and he spent six months in jail. Soon after he was released, she was beaten again but the husband was never arrested again. The man [Vithu] was violent and viciously attacked her more than twenty times. But nothing was done. And so she lost her life. The authority failed miserably in this case. I still feel very sad about this incident because we had spent so much time trying to help both of them.

Vithu had been released from prison without follow-up supervision to monitor his behaviour. He restarted his campaign of violence against his wife, Roumjong, soon after returning to the family home. One day Roumjong went fishing in a local pond to feed her family. There he *bludgeoned* her to death with a large piece of wood. Roumjong's murder, and the other examples I have provided in this section, show how local reconciliation needs to be considered a commissioned act of omission that is repeated over time and can be fatal. Not only this, but women's murdered bodies are an intimate part of crisis ordinary Cambodia in which officials are able to ignore the bloody consequences of their authoritative acts and the pain that those acts produce.[9]

The Law and Lawfare of Forced Eviction

The Pain from Their Beatings Lingers on

Lament the Khmer people crying everywhere, the evil people show no compassion
Oh, the sadness of the people is enormous
We don't know when the end of our karma will be
We used to live in Boeung Kak lake community peacefully
The evil company, Shukaku Inc has violated our rights without thinking
They pump the sand to kill us
Scars from the beatings remain painful
They beat us brutally without thinking about our women or children

They detain us under the pretext of breaking the law
Oh, the powerful and high rank officers, please take a look at us, we are living in sadness
everyday day
We do not know who can help the poor to solve these problems.

As the BKL women's song above makes lyrically clear, forced evictions are rights infringements and legal violations that have led to the literal and metaphorical scarring of women's homes and bodies. BKL has been targeted by the pump, and other apparatuses, for killing. Fatal wounds are encountered, therefore, not just by domestic victims, but also their BKL counterparts. The song brings out the thematic of wounded bodies (the beating and lingering pain) and the geographical body (caught in a geopolitical web of Sino-Cambodian relations via Shukaku Inc.). In this final empirical-led part of the book, I hone my analysis on the outcomes of BKL women's public protest and actions taken to deter them, first, via state-sponsored violence, second, the breakdown of marriages, and, third, through imprisonment. Home dwellers who have already made sacrifices to claim their rights through the subversion of elite power (as seen in Chapter 5) are subject to a government seeking to extend its power and influence over the intimate through the violence of law and lawfare to deter them. These are structures of violence that are supported and mobilised through the use or threat of physical violence, from intimate partners in the home to authorities who instruct and enact violence against women on the street.

State-Sponsored Violence Against Female Activists and the Politics of Omission

> Forced eviction is a form of violence used by the government. Although the government does not beat us directly, they use the armed police and intimidate us physically and psychologically.

As asserted by a female interviewee, BKL women have encountered violences of commission in the form of state-sponsored violence against them as the first consequence of their protest.[10] These encounters go against Article 39 of the Cambodian Constitution, which claims that 'Khmer citizens shall have the right to denounce, make complaints or file claims against any breach of the law by state and social organs or by members of organs committed during the course of their duties'. They also go against purported promises of liberal peace and transitional justice, Alexandra Kent (2016, p. 7) argues, given 'that violence and intimidation are being used to silence people who peacefully demonstrate against an economic and political order they deem unjust'.

Despite assumptions made that BKL women were less likely to be beaten up than their male counterparts, across the long duration of their activism they have been variously pinned to the wall by police and private security guards; beaten with electric batons and shields; hit by slingshots, marble shooters, and sticks;

dragged into police wagons; grabbed by the neck; knocked unconscious; had limbs broken; been pushed to the ground by water canon; and suffered miscarriages after being hit and kicked with lethal force.[11] As noted, these acts of violence are perpetrated by 'armed police'. Otherwise known as auxiliaries, para police, or district security guards, they are 'complicit in numerous human rights violations, including violent assaults resulting in serious injury' across Cambodia (Amnesty International 2015, p. 59). Sitting together with Nget Khun, a 78-year-old activist known as 'Mummy', Srei Leap told me of one particularly memorable instance:

> I clearly recall that Mummy fainted and was paralysed by an electric shock. And at least four to five of our protesters also fainted, were shocked by the electric prod, and parts of their bodies turned blue. We can still see the scars now.

Nget Khun's experiences and those of her BKL peers speak clearly to the intimate war of direct violence and lawfare being waged against them. Violence against BKL women are extra-judicial acts that are part and parcel of the wider security landscape in Cambodia in which capital and political power have been harnessed to merge state and private security into a hybrid security actor (Sidaway et al. 2014). Typically wearing dark blue uniforms, holding truncheons, and hiding their identities behind black motorbike helmets (Figure 6.2), security force personnel have been mobilised against demonstrators with impunity (Amnesty International 2015) – the anonymous security guards dually hiding the anonymity of those who commissioned them.

Figure 6.2   Security guards at a protest in Phnom Penh, 2014. Source: Courtesy of Neil Loughlin.

In 2015, twenty years after the Fourth World Conference on Women[12], Mu Sochua (member of parliament for the CNRP before its dissolution) took the opportunity in the *Phnom Penh Post* to set out what she saw as the achievements of, but also challenges for, Cambodian women. She writes that:

> The state itself commits violence against women who join public protests. Hired security guards and military police well equipped with batons, shields, weapons and tear gas use violent measures to ensure 'social order'. Women, children and even the elderly do not escape this draconian conduct by state forces. And none of the complaints led by the victims has ever been examined by the courts!

In Mu Sochua's candid words, it is possible to see how the act of writing the words of law are quite distinct from the inscription of those words on the bodies of citizens. Therefore, while the government is a signatory of CEDAW, and they have had a ratified DV law for over a decade, this does not mean that women will escape state violence or be in the position to access justice via the courts. Rather, 'social order' concerns become a speech act that the government utters to reconstitute women's peaceful activism as a national security threat. This positioning of BKL women as dangerous citizens who have failed in their obligations to conform to the prevailing social order both licenses the commission of violence against them and the omission of action taken in response to these legal violations.

The treatment of Bov Srey Sras, an active member of the BKL community, attests in very personal terms to these observed dynamics and their deadly combination. Bov Srey Sras suffered a miscarriage in 2012 after a police officer drove his baton into her stomach and then kicked her at a protest. She submitted a complaint to the Phnom Penh Municipal Court yet no action was taken, despite video evidence. Instead, Deputy Phnom Penh police chief Phoung Malay (cited in Worrell and Chakrya 2013, np) responded:

> Is the victim old or young, and does she sue me to return her kid? I want to tell her that if she wants to get back her kid, I am also young.

As the Cambodian Center for Human Rights (CCHR 2013) observed, '[h]is comment made to *The Phnom Penh Post* was widely interpreted as suggesting that he could impregnate her as compensation'. In response to the suggestion that she might like to have sex with him to be impregnated and replace the child, Bov Srey Sras replied (cited in Chakrya 2012, np), 'For me, what Phoung Malay said is an insult to all Boeung Kak women ... And it is unethical behaviour that is insulting to women nationwide.' Not only had Bov Srey Sras suffered a miscarriage commissioned by state organs, but she also contended with violences of omission arising from impunity. The female director of a Cambodian human rights NGO also noted this issue in an interview with me:

> All the violence that women have encountered at Boeung Kak was carried out by the authority, and the authority is under the government. We do not say that the Ministry of Women's Affairs did that, but the government. But the Ministry does not take any action to protect the women ....

The suggestion being made in the above interview is that MOWA are blame-worthy in their silence to publicly denounce, or take action on, violence against BKL women. They too are imagined complicit in violences of omission that tac-itly accept and thereby perpetuate the status quo. This sentiment was also shared by one BKL activist with me:

> We know that the Ministry of Women's Affairs is always campaigning to stop domestic violence. It paints this picture to avoid responsibility while the government is harm-ing women. Instead we want the Ministry of Women's Affairs to send the message to the wider government to stop violence against women everywhere. This is for all women. If we think of domestic violence, it isn't as bad as violence by public author-ities who very well understand laws on women's and citizens' rights. Because public authorities have a role to educate and advise husbands and wives to stop violence, they should be the model for citizens. In contrast, they are now teaching others to use violence publicly. So far, we see violence along the road, in front of institutions when we are demonstrating for justice. Public authorities use force to hit us on the head and some women lose their babies in the womb. Actually, these public authorities should be punished more severely than others because they know the law well ... the Ministry of Women's Affairs should be safeguarding us. We are protecting our homes in a peaceful manner. They should take action to pressure or to file a charge against those police officers that harm us. There is no possibility of stopping domestic vio-lence while violence in public is ongoing. The situation only gets worse.

As the BKL activist alludes, dealing with domestic violence is core to the work of MOWA, yet in BKL women's eyes they are complicit in the silence and impunity surrounding the violence women have encountered on the street. The interviewee claims that it 'paints this picture' of care through subterfuge and fails to safeguard the safety of 'all women'. Her analysis of this supposed behaviour can be elaborated upon further by looking at Geraldine Pratt's (2005, p. 1056) work on 'legal abandonment', which occurs 'through a complex and gendered layering and enfolding of geographies of public and private, one into the other'. In the context of research on murdered sex workers and live-in carers on tempo-rary work visas in Canada, she posits (drawing on and critiquing Agamben's 1998 work) that there exists an 'uncanny capacity on the part of the state to regulate and police certain types of violence and illegal behaviour' (p. 1052). Pratt (2005, p. 1053) continues:

> I want to pursue this issue of absences and lapses in state policing and regulation in particular spaces of the city. Rather than viewing such lapses as aberrations from normal practice, I want to ask how such irregularities become the norm for certain people in certain places.

Following this line of argument, it could be argued that abandonment, or vio-lences of omission, are not exceptional but rather routine for Cambodian women like those from BKL. States of 'bare life' (Agamben 1998) can arise for people with legal protections of citizenship (Ramsay 2019a), but who because of their obstruction of profit-making results in their precarity and abandonment by state systems of support.

Firefighting these violences of omission have, in turn, been seared into the crisis ordinary. Dovetailing with this point, Tep Vanny widened such charges of omission still further at a march organised to deliver a petition calling for the end of government brutality against housing activists in November 2013. She decried that:

> UN Women has based an office in Cambodia, but they have never paid attention to all the violence committed by the armed forces during forced land evictions and in garment factory work (cited in Narim and Willemyns 2013).[13]

The 500-person strong, mainly female march, was led by Yorm Bopha; it was organized to coincide with the annual 16 days of Activism against Gender-Based Violence campaign and the International Day for the Elimination of Violence against Women,[14] which annually falls during this period on 25 November. BKL women thus used this, and other dedicated days, to widen the remit of their cause and claims (as seen in Chapter 5). These were long-standing global initiatives in the violence against women arena that BKL women felt their experiences of gender-based violence belonged within. Yet many felt they were usually only intended for certain women suffering certain types of violence. Jenna Holliday, then spokesperson for UN Women in Cambodia, responded to the critique by telling reporters that it had published an opinion piece in the preceding week that demonstrated their awareness of violence against BKL women and the need to raise this with the government (UNWOMEN 2013). The designated days for action were, once again, being used as punctual spaces to visibilise their cause, only now their rights claims were being articulated during, and strategically morphed into, wider calls to address gender-based violence. In contrast to the standard inclusion of domestic violence, BKL women were challenging the orthodoxy of their exclusion in these public events.

The *Cambodia Gender Assessment* (MOWA 2014b), which has a dedicated chapter on violence against women (and girls), similarly omits reference to the violence that female human rights defenders face.[15] The government-led publication covers (in order): domestic violence, sexual violence, rape, gang rape, sexual harassment in workplaces and the community, and sexual exploitation of women and girls. Other policy instruments are also culpable to criticism for their omissions. The National Action Plan to Prevent Violence Against Women 2014–2018 (NAPVAW 2) (MOWA 2014c) spotlights violence against women with increased risk as one of three priority violence issues and does so with specific reference to women with disabilities (see Astbury and Walji 2014), women living with HIV, and sex workers. However, while the government cites commitment to a broader range of victims' voices and their needs in recent reports and strategies, the prioritisation of 'violence against women with increased risk' remains discriminatory. In other words, certain types of violence and wounding against women can, and will, be silenced by officialdom when it is politically expedient to do so (Brickell 2017a). The violence that BKL women have encountered is archetypal here and speaks to their uninvited status as previously evidenced in Chapter 5.

Although BKL women had not always framed their rights violations as ones tied to the global arena of violence against women, such 'rights talk' (Merry 2003) has become more evident over time and in response to these observed exclusions. For example, during my February 2014 visit to the BKL advocacy centre, I found many of the women (including Phan Chhunreth in Figure 6.3) wearing a 'Stop Violence Against Women' T-shirt.

Figure 6.3   Phan Chhunreth in a 'Stop Violence Against Women' T-shirt, 2014. Source: Katherine Brickell.

The T-shirts were being worn in advance of International Women's Day on 8 March to recognize the violence they and other women nationwide were enduring. As one interviewee explained:

We have this slogan because we have seen Cambodian government and authorities abuse Cambodian women cruelly. Every day, we are hit and threatened the minute we are out protesting.

The right of women to be free from violence was used both as an ideal and as a tactic for BKL women to make their experiences of violence 'count' and to make visible their politicised omission from policy discourse and planning by domestic and international actors. They had co-opted such rights talk as a form of 'rightful resistance' (O'Brien and Li 2006) to press their claims for recognition and moral accountability as victims of violence against women. That 'human rights talk is the most popularly recognizable feature of a non-existent, or at least severely compromised system of rule of law' (Wilson 2007, p. 360) is therefore particularly apt in the Cambodian context. This has been demonstrated across the domestic violence and forced eviction case studies.

Forced Eviction, Marital Breakdown, and Domestic Violence

> I am really hurt. Outside the house, the authorities beat me; inside the house, my husband abuses me, so it is almost an unbearable pain. At some point, I want to escape from the house, and let him live alone! On the other hand, I feel sympathy on my children. If I did not have children, I would have run away and given my husband the house! This pain almost kills me, and it has been so extreme that I cannot stand it (Phan Chhunreth).

In this section of the chapter I frame marriage and women's bodily integrity as the collateral damage of external state relations that have expedited forced eviction in BKL. Although it cannot be said that every marriage or every home was necessarily a happy one prior to their protests, women's perceptions are that marital strain and/or breakdown has become a more commonplace trauma.[16] Post-eviction, its detriment continues to be lived through truncated and violent marriages, which are having a lasting legacy for participants in my study. The intimate wars that many BKL women are dealing with are thus multiple and mutually imbricated. Blame for them lies geographically close to, but also far away from, the home as the locus point of proximate and distant relations.

As Phan Chhunreth explains, for many women experiences of violence in the public sphere have been compounded by exposure to domestic violence emerging out of their activism against forced eviction; their combination is so acutely felt that she describes the pain as tantamount to being killed.[17] As such, numerous BKL women face violence in both public and private domains of contemporary Cambodia and demonstrate how the issues of violence against women and forced eviction intersect in a toxic concoction of intimate war. In *'Good Wives': Women Land Campaigners and the Impact of Human Rights Activism*, LICADHO (2014b) reports, for example, that almost all of the 24 women interviewed said that their new roles and the resulting changes in their relationships with their husbands had led to arguments. Five women also said that these arguments had led to violence, which they saw as a direct result of their role as campaigners. That none of the women had suffered from spousal violence before they began their activism is an important rejoinder to the idea that forced eviction is part of the 'new war' being waged against women (Federici 2018b). This evidence again renders their out-of-frame status particularly problematic when it comes to action on violence against women in Cambodia. A 2014 interview with the female director of a human rights NGO in Cambodia made this additionally clear:

> Women suffer twice. First of all, when there is the forced eviction, the women are the people who come out to peacefully struggle and challenge with the authority in the absence of their husbands who go to work. When the authority evicts them, they say they will be compensated. The authority provides compensation by transporting them to a relocation site which is far away from the city. Husbands have to stay in the city otherwise they lose their job; for example, they are a *motordop* [motorbike] driver and need to earn a daily income. Men therefore stay there to earn money in order to support the family. However, after a while, when the transport becomes expensive, the husband does not go back and has another family. As a result, the women have suffered [a] second time in addition to the first suffering. Upon the forced eviction, domestic violence also starts to happen in the family. So, eviction has resulted in a negative impact against women who suffer more than the men. Women carry a very heavy burden on their shoulders.

While in Chapter 5 I indicated that men in the community had been largely in favour of women's leadership role and public representation of family interests, spousal tensions have nevertheless arisen that demonstrate the multiple and coexisting battlefields that women negotiate under the everyday burden of the crisis ordinary. This is something that Tep Vanny felt keenly:

> I used to be a simple housewife, but immediately I had to turn myself into an activist who confronts authority …. I stand in a very dangerous position. This instant shift has deeply influenced my emotions and thinking. And now I have to deal with problems in the family! Before, it was fine although my husband and I had arguments. Now, even though it is a little argument, it turns very big … it is caused from a mixture of mental problems, and one of their root causes is from this development.

Although Tep Vanny's husband had taken on a greater share of the housework and adopted a more flexible attitude to his gendered responsibilities in the home, this did not preclude growing resentment and tension between the couple. Tep Vanny's interview explains how small arguments had escalated to larger ones and were tied, in part, to the mental stresses and strains of her activism, which eventually resulted in divorce. Their marriage had also come under government pressure on account of his position as a high-ranking military officer. Tep Vanny's then husband faced a range of threats and an investigation was ordered by a military court. In such situations then, 'displacement goes beyond a change in residential landscape, often resulting in the destruction of the household unit' (Daley 1991, p. 249).

Many BKL women discussed their gendered duties with me as an evolution rather than transgression of their domestic roles (see Chapter 5). Yet this did not mitigate difficulties of combining protest with existing domestic duties such as housework. For some men too, BKL wives' inability to fulfil both their private and public commitments as housewives was considered the source of disobedience by husbands who lost patience over time. Srei Leap, the youngest and then unmarried member of the group explained:

> Well, there are family problems like arguing within the family. Because women spend their time protesting, we leave our children at home, we do not have time to cook. After our husbands come back from work, they sometimes feel angry about this unpreparedness in the family, so there are arguments. Several families have decided to divorce.

BKL women were caught in a double bind: either they step into the uncharted territory of public activism and risk divorce and even spousal abuse or else they leave their dispossession uncontested and lose the home upon which their marital life is built. The difficulties that Srei Leap mentions above, of maintaining home life in combination with advocacy work to ensure they can provide 'shade', links to pressures felt in other South East Asian countries. In respect to Thai labour activism, Mary Beth Mills (2005, p. 133) details how women's protest was acknowledged as a 'conflict with gender- and kin-based obligations'. In a similar vein, Srei Leap made named reference in her interview to Bo Chhorvy whose marriage ended on account of her public activism and supposed dereliction of domestic duty. Going one step further, Bo Chhorvy explained how this conflict was in fact *favoured* by government authorities:

> I am divorced because I go out to protest a lot ... I do not have time to take care of housework, like cooking, and my husband disagreed with my protest. He wanted to keep me at home, but I opposed him. Therefore, we divorced ... everyday, the authority prays for us to have problems at home like this, so that we have less time to protest against them about the land dispute. The authority hates my activism, and it is good for them to see me having family problems, they want revenge on me.

The insinuation being made by Bo Chhorvy is that marital strife is not just collateral damage of forced eviction but is also instrumental to a government intent on undermining the ability of BKL women to sustain their activism. Suffering becomes a 'political technology' (Davies et al. 2017, p. 6) to quash dissent. This state of injury is sought not only through state-sponsored violence and detention against BKL women but also through familial disruption. This manifests both in domestic violence and the inability to seek help and justice in response. As a LICADHO (2014b, p. 9) report makes clear, BKL women and other activists find themselves in a compromised position when it comes to seeking help from any kind of authority:

> Firstly, because they are land activists, the authorities regard them as troublemakers and are therefore unlikely to make any efforts to help them. Those activists with a particularly high profile said that they feared that if the authorities found out about the domestic violence they would not only refuse to help them but would use the information to discredit them. One of them said that her local police often follow her and monitor what she does so she has no faith that they would help her. One other woman, who experienced serious violence in the family, said that she did eventually ask the police for help but they told her it was an internal family matter and took no action to help her. None of the other women interviewed had approached the authorities for help.

Given these omissions, what is at stake on the 'Stop Violence Against Women' T-shirts and their message is reconnecting and legitimating their struggle within the normative remit of women's rights talk and domestic violence concerns. BKL women's aim is to reframe their activist citizenship as acceptable rather than aberrational and switch up their 'troublemaker' status. Nevertheless, the prospect of seeking legal redress from authorities for the domestic violence they faced was feared to be a false move. Women were suspicious that communicating 'fire in the house' too publicly would be fruitless or else would be knowledge weaponized to wreak extra harm against them. BKL women thus performed given as well as invented citizenship scripts in different circumstances to advance their goals at the same time as trying to protect themselves from further scrutiny and harm. Their non-reporting speaks to the gender-based discrimination that women human rights defenders often experience as they have less access to existing protection mechanisms (Office of the High Commissioner for Human Rights (OHCHR) 2017).

The regular media reporting, particularly in the government-controlled Khmer press, on the state of BKL activists' marriages supports the point that women's marital circumstances were being strategically mobilised to discredit them. In January 2014, for example, it was reported that having been recently released from prison, Yorm Bopha was locked up and abused in her home by her husband (Odom and Blomberg 2014). According to a number of government-aligned news outlets, this was supposedly because she had committed adultery with the village representative and had not ceased her activism as per her husband's wishes. Yet as a BKL community member told *The Cambodia Daily*, Yorm Bopha's husband had become very close with local authorities – a wedge that authorities were driving between spouses to suppress future protests (Odom and Blomberg 2014). Such stories show, Alexandra Kent (2016, p. 21) reflects, 'how conflicting interests over property in Cambodian society may be driven off-stage into the unofficial, domestic realm of personal relationships'. Described as a 'divide and conquer' tactic by influencing Yorm Bopha's husband (IFEX 2014), the couple went on to divorce in October 2014. She told reporters:

> After I came out of prison last year, my husband and I separated. He was gambling and drinking and he gambled away a lot of our property from the house. He was also violent towards me, so I couldn't stay with him. Domestic violence is a huge problem in Cambodia, and I felt like I had to set an example. But my husband took our son with him, and I only see him once a week, which is really hard. I feel like it's all been so unfair – I spent one year in prison, and then my family fell apart. But women will always face obstacles in life, no matter how big or small, and we have to be strong. All these obstacles will only make me stronger (Yorm Bopha cited in Wight and Muong, 2014, np).

As Yorm Bopha makes clear, the breakdown of her marriage 'set an example' that women will not stand for domestic violence and that the multiple obstacles, big and small, put in their way may embolden rather than discourage their independent decision-making and desire to protest.

The Lawfare of Incarceration

The third outcome of BKL women's protest opposing the gendered insecurities of forced eviction has been that lawfare has been legitimated against the grass-roots network. This has been achieved via imprisonment and the de facto criminalization of peaceful protest (see Box 6.1 for a timeline). Many forced eviction activists I interviewed have therefore dealt with multiple displacements from home and family life.

As Box 6.1 captures, there have been four substantive periods of incarceration for different cohorts of the group and individuals. In their report, *Courts of Injustice: Suppressing Activism through the Criminal Justice System*, LICADHO (2017, p. 5) explores the mechanics of harassment through the criminal justice system and the procedural tactics adopted: the use of unsubstantiated criminal charges, failures to comply with international fair trial standards, and how the arrests, trials, and convictions have been timed to coincide with important political events. These legal moves are ones that Human Rights Watch (2015, pp. 38 and 72) note specifically in relation to BKL women's experiences:

> Cambodian authorities have repeatedly subjected community members protesting Bank-financed projects to arbitrary arrest or trumped-up legal actions aimed at preventing them from protesting .... Those charged have been routinely denied bail, convicted after expedited and truncated trials that did not meet international standards and did not give the accused adequate time to prepare and put forward a defense, and given significant prison sentences.

Box 6.1   Timeline of BKL women's arrest, incarceration, and release

24 May 2012

Arrest of Nget Khun, Tep Vanny, Kong Chantha, Song Srei Leap, Tho Davy, Chan Navy, Ngoun Kimlang, Bov Sor Phea, Cheng Leap, Soung Samai, Phan Chhunreth, Heng Mom, Toul Srei Pov ('BKL 13').[18]

26 May 2012

Conviction of 'BKL 13' under Articles 34 and 259 of Land Law for illegal occupancy of public property and Article 504 of the Penal Code for obstructing public officials with aggravating circumstances. All of the women sentenced to 30 months in prison, but 6 of the women receive suspended parts of their sentences.

27 June 2012

Release of the 'BKL 13'.

27 December 2012

Conviction of Yorm Bopha by the Phnom Penh Municipal Court of 'intentional violence with aggravating circumstances' for ordering an attack on two tuk-tuk drivers and is sentenced to a 3-year jail sentence.

22 November 2013

Release of Yorm Bopha on bail.

10 November 2014

Arrest of Tep Vanny, Kong Chantha, Nget Khun, Song Sreyleap, Phan Chhunreth, Bo Chorvy, Nong Sreng ('BKL 7') during a peaceful protest in front of Phnom Penh's City Hall to draw attention to the repeated flooding of their homes.

11 November 2014

Conviction of 'BKL 7' under Article 78 of the Traffic Law for obstructing traffic by the presiding judge.

11 April 2015

Release of 'BKL 7' by royal pardon. The release of the women comes as part of recent political deals between the ruling and opposition party leaders.

17 August 2016

Charging of Tep Vanny and Bov Sophea with incitement following a peaceful demonstration during week 15 of the 'Black Monday' protests.[19]

23 February 2017

Conviction of Tep Vanny for 'intentional violence with aggravated circumstances' against a security guard (Article 218 of the Penal Code) and sentencing to 30 months in prison for peaceful protest in front of Prime Minister Hun Sen's house in 2013.

20 August 2018

Release of Tep Vanny by royal pardon.

The imprisonment of the 'BKL 7' in November 2014 is a good example of these observed dynamics. Each were sentenced to a year in prison, the maximum allowed by the law, and fined US$500. These sentences were given in a summary trial lasting less than three hours and held just twenty-four hours after their arrest for blocking a main road outside Phnom Penh's City Hall with a bed frame (Human Rights Watch 2014). Although such offenses would normally attract no more than a small fine or simply a warning, the 'BKL 7' had been subject to the art of lawfare in which 'veils of legality' (Hoffman and Duschinski 2014, p. 506) were used to legitimise their incarceration.

The disproportionate and politically motivated use of the Traffic Law to curtail their activism jars against the impotency ordinarily associated with law in Cambodia (identified in Chapter 5). Sebastian Strangio (2014, p. 166) explains that:

> What laws do exist are enforced so selectively that they might as well not exist. Traffic police, crouched behind parked cars on busy thoroughfares, jump out to shake down unsuspecting motorists for bribes, whether or not they've committed an offense, while vehicles with military and state license plates breeze past unmolested.

On 24 June 2016 Hun Sen entered the performative foray of the Traffic Law. Revealed in the Cambodian media, the prime minister had been fined 15,000 riels (around US$3.75) for riding a motorbike without a helmet (*Khmer Times* 2016). The police officer was quoted as saying that, 'I have a role to enforce the traffic law for everyone properly. So whether you are powerful, poor, rich, small or high-ranking, if you break the law, you must be punished the same by the law' (Asian Correspondent 2016, np). On his Facebook page, Hun Sen said that he accepted the fine, adding he would personally pay for it. He responded, 'I appreciate Koh Kong province's Srae Ampel district police who implement the law without discrimination and with independence and without any fear of powerful people, including the prime minister.' In this instance, his performative gesture was being used as a vehicle to demonstrate legal compliance and speak more broadly to the supposed independence of the judiciary in Cambodia. Yet the incarceration of the 'BK7' through lawfare makes a mockery of these claims.

Given repeated imprisonments, BKL women's activism increasingly turned to advocating for their freedom. Writing about environmental activists Wangari Maathai and Rachel Carson, Rob Nixon (2011, p. 145) argues that their 'marginality was wounding but emboldening, the engine of their originality'. In a similar way, this could also speak for BKL women under the weight of arrests. In February 2013, for example, BKL women held a protest outside the US Embassy where a press conference was being held by (then) Ambassador William Todd on human rights issues and to which two BKL women, Tep Vanny and Heng Mom, were invited. The pictures of the two US leaders, Barack Obama and Hillary Clinton, were held up by BKL women in thanks but were also used to lobby for further pressure on the Cambodian government to free Yorm Bopha. It is widely believed that Yorm Bopha had been targeted as a result of her activism and outspokenness, especially during the campaign for the release of the 'BKL 13' (CCHR cited in IFEX 2013). She had previously been warned by a police officer that she was on a blacklist on account of her new-found predominance and was subject to regular threats, intimidation, and harassment (LICADHO 2012). Yorm Bopha had been convicted on bogus grounds for a planned attack on two motorbike taxi drivers.

On posters at the protest the message in Khmer read 'Stop using the court system to arrest land activists' and on their heads group members wore photographs of Yorm Bopha's face with the words 'Free Bopha' as they spoke to the attending media (Figure 6.4).

Figure 6.4   Tep Vanny interview with the media outside the US Embassy in Phnom Penh, 2014. Source: Katherine Brickell.

BKL women also moved once again to evoke the power of mother activism I discussed in Chapter 5 to press their claims for release. The imprisonment of Yorm Bopha led to other imaginative protests which spoke out against the intimacy between mother and child denied. Women and children wore hats, for example, covered by bird nests with eggs nestled within them. Some of the very reasons that DV victims have become reconciled to violence, in order to mother within an 'intact' marriage and home, were therefore experienced rather differently for BKL activists who demonstrated their loss of habitat and capacity to mother. Srei Leap told me more:

The idea of putting the eggs in the nest on our heads was to express our need as humans for a house to live in. This picture [pointing to a photo book] shows the demonstration where we protested to free Yorm Bopha who has been sentenced to imprisonment … as the result, her children are taken care of by their father alone. We compared the warmed eggs sat on by their hens to Yorm Bopha's children who do not have the opportunity to have their mother close by.

The protest connects to Ananya Roy's (2017, p. A4) recent work on anti-eviction protest and emplacement – 'the process or state of setting something in place or being set in place' – through intimate practices of constructing domes-ticity. BKL women had curated their domesticity in the public sphere through the nest and eggs to make spatial claims to the homes removed from them and their children; a 'creative defense against resignation to a bad world' (Berlant 2014, p. 27). The mother trope is also one that Amnesty International played with in their advocacy work for Yorm Bopha, named prisoner of conscience by the orga-nisation. One of their campaigning materials that circulated online included a photo of tearful Yorm Bopha in prison uniform holding her son close. Alongside, there were four tick boxes, the first, activist, the second, wife, the third, mother, and the fourth, criminal. The fourth box is crossed out and it reads underneath 'Yorm Bopha does not belong in jail'.

With the multiple waves of arrests and periods of incarceration they were subjected to, the women once again escalated their local struggles to ones calling for transnational solidarity. As such, carceral landscapes, including those in Cambodia, 'can be, and have been, productive sites for the formation of new alliances and solidarities that contest and subvert legal constitutivities and the networks of power it creates and maintains' (Villanueva 2017, p. 17). Following realist and neoliberal traditions of international relations, which suggest that states only comply with the principles of (international) law when it is in their national interest (Hafner-Burton and Tsutsui 2005), the Cambodian government used the release of Yorm Bopha as a diplomatic means to perform compliance and avoid potential embarrassment at a politically sensitive time; the fates of BKL once again were caught up in eventful geographies above them.

The incarceration of the 'BKL 13' prior to Yorm Bopha's jailing is also exem-plary here, having garnered the sympathy and support of then US Secretary of State, Hillary Clinton. In mid-June 2012, Hillary Clinton met with Foreign Minister Hor Namhong in Washington and passed on her encouragement for the 'BKL 13' to be released (Di Certo 2012). According to *The Phnom Penh Post* (Jackson and Turton 2016), emails leaked by Wikileaks confirm the reassurances given to Clinton at this time that their release would be made. Set to be an inter-national embarrassment, given Cambodia's then chairing of the Association of South East Asian Nations (ASEAN), the women were set free by a Cambodian appeals court on 17 June (Figure 6.5).[20]

Figure 6.5   Release of the 'BKL 13', 2013. Source: Courtesy of Nic Axelrod.

When I revisited in February 2013, photographs of Hillary Clinton adorned the walls of the women's advocacy centre and included children's drawings of her stating their thanks for her intervention (Figure 6.6).

Figure 6.6   Drawing of 'Mother Tep Vanny' (top) with 'Nana' (middle) and a boy (bottom), 2013. Source: Katherine Brickell.

The veneration of Clinton echoed writing on her previous visits abroad as First Lady of the United States (1993–2001) in which hallmark agendas of her visits were women's rights, education, and democratic political systems (Kaplan 2001).[21] Michelle Obama was also called upon by BKL women and their children to raise the visibility of their claims to release when she was First Lady of the United States (2009–2017). In March 2015, children of the 'BKL 7' gathered in front of the US Embassy. They submitted an open letter calling for intervention from Michelle Obama to support the release of their mothers during her official visit to Cambodia to promote education for girls (Free the 15 2015). Drawings representing their mothers in prison with handwritten personalised messages were also submitted to the Embassy with the hope that Obama would read them. The young children clutched family photos of the presidential family and then shouted: 'Please, Michelle Obama. Help our mothers.' Held also by adult protestors (Figure 6.7), these images foreground the emptiness of the mothers' presence. They asked of Michelle Obama, and her audiences, how would *you* and *your* daughters feel in this situation?

Figure 6.7   Reaching out to Michelle Obama outside the US Embassy, 2015. Source: Courtesy of Radio Free Asia.

Often referred to with the moniker 'Mom-in-Chief' and framed as 'mother' to the children of the nation through her policy choices (Hayden 2017), BKL women also positioned Obama as a means to stage their own transnational escalation of their cause and claims to motherhood denied. Family and motherhood, and their adverse exclusion from it, was thus mobilised for strategic ends to deal with the intimate war they were battling.

The Cambodian government have continually tried to delegitimise BKL women as citizens and as mothers to undermine their rights-based activism and the affective potency of their demands, but BKL women have relentlessly pushed back. They have shown themselves to be strong and defiant subjects who refuse to be rendered docile by a government willing to enact the violence of incarceration upon them. Lawfare thereby followed the 'grammar of security' (Gordon 2014, p. 326), which constructs a higher-order plot that BKL women, and especially Tep Vanny, were an existential threat to social order. As Kevin Doyle, ex-editor of *The Cambodia Daily* (2004–2014), told me in 2018 of the group:

> Whereas other leaders of peaceful collective movements in Cambodia suffered instant and complete violent removal through assassination – Chea Vichea, Chhut Wutty, Kem Ley[22] – and, arguably, were targeted as symbols and as warnings to others against collective action that challenges entrenched political and economic interests – the Boeung Kak ladies defied that reality and subverted narratives of fear and acquiescence. The Boeung Kak protesters were more than demonstrators, or smart protest performers, they became the embodiment of the political in Cambodia. Hence, their near total erasure through incarceration. Other individuals who have embodied the collective experience of injustice in Cambodia have not been as fortunate, as their 'erasure' has been lethal and final.

While the commissioned incarceration of BKL women has become a predicable mode of trying to intimidate them, the point being made by the ex-editor is that these enforced 'erasures' from public life have not been made permanent through death like others allegedly have.[23] While suicides and miscarriages in the community tell a different story of lives lost, it has been incarceration that has become the most regularised form of intimate war. Although incarceration did not deter everyone from future activism, core members such as Srei Pov struggled with the adverse physical and mental health effects of her sustained activism and time in jail. Srei Pov's interview with me in March 2013 (she was released from prison in June 2012) also spoke to the wearing down of her body over time and the need for change:

> I decided to end my participation in the community because after I was freed from the prison, I became sick … my health was getting worse when I was sentenced to imprisonment, I was mentally and physically exhausted … these five years, I never had a full sleep. Some days, I am awake until 2 a.m., 3 a.m., or 4 a.m. I worry what tomorrow will bring or lose. There have been too many things to focus on. It almost leads me to mental disorder. In the meanwhile, the psychiatrist proscribes me medicine to fall asleep. Yet, I stay awake some nights and have to overdose on the medicine to sleep.

As Srei Pov's interview, and the song title 'The Pain from Their Beatings Lingers On' testifies, the fatigue and anxiety of activism and incarceration had a deep hold in the psyche and continued to have an attritional force. Just as Peter Kraftl and

Peter Horton (2008, p. 511) argue for the need to take sleep more seriously, a prompt 'located in a renewed concern with geographies of health and physical and emotional well-being', women's testimonies, much like those of domestic violence victims in Chapter 4, show sleep denial to be a daily form of intimate war. While for 'activists, the ambition to survive the world and further disturb it produces psychic loads so very difficult to carry' (Berlant 2018, np), Srei Pov had come to the realisation that she needed to seek out personal breathing room.

The reasons behind Srei Pov's withdrawal from the protests were also rooted in the breakdown of her relationship with Tep Vanny; their intimacy and friendship becoming one of the many casualties on the battlefield of eviction. The accolades that Tep Vanny received (detailed in Chapter 5) undermined group cohesion. At the time of the Vital Voices award, I spoke with multiple NGOs in Cambodia who expressed concern that the accolade, given to a singular representative of the group, was divisive and problematic. Independent and state media picked up on the growing discord between the pair. As Tep Vanny noted, 'I think this is a trick of authorities … when we are demanding justice in a country with no transparency and there is no respect for human rights like this [the government] always has strategies to break up our group' (cited in Geary and Narim 2013, np). Months after the award in July 2013, a number of stations, seen as heavily aligned with the government, also broadcast criticism of Tep Vanny, including lengthy excerpts of a press conference led by her former cellmates in Prey Sar prison, Srei Pov and Heng Mom (Chakrya and Worrell 2013). While the splintering of the group was of clear strategic advantage to the cause's detractors, internal rifts and pressures had arisen for some time in the leadership of the group and the intense friendship between Tep Vanny and Srei Pov.

The documentary film *A Cambodian Spring* (Kelly 2018) gives some sense of this. Srei Pov criticises Tep Vanny for being hungry for fame and spreading rumours that she had 'sold out to the government'. At the film's start and then mirrored at its end, she is filmed at home with a wallet of A4 photographs in hand. She flips through the photographs reminiscing: with Tep Vanny in Democracy Park protesting together; relaxing under Om Leng waterfall with other BKL women in happier times; the pair posing together in prom dresses. With tears welling up in her eyes and her head twisting down in sad contemplation, she says:

> If we have unity, compassion, and trust, then we will be strong, and no one will break us. But if we don't trust each other then how can we work together? It will all come to an end. We won't succeed.

While Srei Pov dropped out from her activism to focus on her family and home life, Tep Vanny's notoriety and leadership only grew but also led to her own retreat from public view. From August 2016 Tep Vanny was held, for the first time alone, in the notorious Prey Sar jail. She was found guilty in February 2017 under Article 218 of the Cambodian Criminal Code (Royal Government of

Cambodia (RGC) 2007b) for allegedly assaulting Hun Sen's security guards. The charges originally brought against her relate to her participation in a peaceful protest in front of Prime Minister Hun Sen's house in 2013, during which she and other activists called for the release of a detained fellow community member. During the trial, no credible evidence was presented to substantiate the charges and no witnesses were presented by the prosecution, thus preventing any cross examination by the defence (Tang and Thul 2017, np). Furthermore, the court refused to hear testimony from witnesses supporting Tep Vanny's account that she and other protesters did not commit any violence during the protest. She shared a concrete 6 ×16 metre cell with 84 other women (Coffey 2018). Specifically in relation to the case, the 2017 report of the Special Rapporteur on Cambodia, Rhona Smith (United Nations General Assembly 2017, p. 12), explains:

> Tep Vanny was prosecuted following her participation in a black Monday event on 15 August 2016. In contrast to other participants, who were detained briefly and then released, she was then charged with 'intentional violence with aggravating circumstances' relating to another protest in 2013. On 23 February 2017, she was sentenced to two and a half years' imprisonment .... Taken together, these incidents and misapplication of laws has had the effect of restricting civil society space and diminishing the perceived democratic space in the country. There has been a negative trend whereby political activists, human rights defenders and journalists continue to face restrictions when exercising human rights and fundamental freedoms.

Tep Vanny's incarceration is perhaps the purest articulation of lawfare against BKL women: 'the resort to legal instruments, to the violence inherent in the law, for political ends' (Comaroff and Comaroff 2009, pp. 36–37). The timing was not incidental with the then upcoming election in July 2018; the ruling government was being particularly paranoid and active in its attempts to break the spirit and destroy the resolve of a community who had become a core part of the 'black Monday' protests that Rhona Smith refers to. Beginning in May 2016 with the arrests of staff from the NGO ADHOC, campaigners against the increasingly repressive politics of the state encouraged the public to wear black T-shirts as an attritional gesturing of weekly dissent.

BKL had, by this point, achieved a powerful platform of resistance against the necrocapitalist forces of contemporary Cambodia with Tep Vanny having become its best known public face (Human Rights Watch 2017b). As her mother comments in a media piece on her turn from housewife to 'grassroots warrior', she has become 'the strongest woman in the country' (cited in Retka and Odom 2017). For this reason, she has been targeted, according to LICADHO's Naly Pilorge, because 'she was perceived as a leader, an orator and activist, and in some instances she is the face of the Boeung Kak lake issue' (cited in Baliga and Chheng 2017). As Phil Robertson of Human Rights Watch corresponds, the 'case against Tep Vanny is a blatant misuse of prosecutorial power to punish her for her peaceful activism .... This prosecution is intended to silence Tep Vanny and intimidate other Cambodian activists.' Her repeated jailing is in many ways an admission that the Cambodian government had found her unmanageable, the embarrassment of this requiring the denial of her liberty and physical presence in public life. As an opposition female politician commented to me in a 2017 interview, 'She's a symbol of strong Cambodian women … claiming not only her rights but fighting for justice in society in general … she's a symbol, so they want to crack down on the symbol.'

After the July 2018 election, which saw the CPP take all 125 parliamentary seats, the US State Department called for the release of all political prisoners and an expanded set of visa restrictions on the country's political elite was applied. The European Union added its willingness to enforce sanctions and threatened to strip Cambodia of its 'Everything But Arms' (EBA) status, which gives the country tax-free access to European textile markets. Organisations like Amnesty International also intensified their global campaign for Tep Vanny's release (Figure 6.8).

Figure 6.8   Amnesty International protest outside the Cambodian Embassy in London, 2018. Source: Courtesy of Amnesty International.

Less than a month after the election, Tep Vanny received a royal pardon, requested by Prime Minister Hun Sen, and was released (Kijewski 2018). Tep Vanny's release was seen as a popularist move to rebuild legitimacy by Hun Sen post-election. Commentators on social media urged Western governments not to be fooled by his smoke and mirror actions following a long-standing pattern of clamping down and then, under a benevolent guise, releasing prisoners at politically apposite times. Global Witness (2018, np) identifies that:

> For those close to them, the release of these brave activists will take some of the sting out of Cambodia's slide into dictatorship. But we mustn't be fooled by small signs of mercy. These people were locked up to send a message – that Cambodians shouldn't dare speak badly of their government or challenge its dominance. Their release is not symptomatic of a change of heart, but a change in strategy.

With victory secured, the CPP-run government were trying to mitigate the widespread international condemnation they faced over the lack of a free and fair vote. The taking and then strategic giving of liberty is a core example of the lawfare brandished against Tep Vanny and other BKL women over the years. Leaving prison, Tep Vanny emerged wearing a jumper she had knitted in prison. At its centre was a Cambodian flag (Figure 6.9).

Figure 6.9   Tep Vanny upon release from jail, 2018. Source: Courtesy of Leonie Kijewski.

Tep Vanny drew on symbolism, and craftivism, to show her citizenship of, and devotion to, Cambodia. Talking to Amnesty International in October 2018, two months after her release from prison, she reflected on the last decade of her activism:

> Imagine having to campaign for the right to live in your own house. Imagine being arrested and imprisoned for that .... I spent two years and five days in prison.

I missed my children, and my mother, who is elderly and sick. They needed me, and it was horrible to think that I couldn't comfort them. I did not know how to live from one day to the next. The only way was to make myself busy with work, and I did a lot of knitting! It helped reduce my stress and stopped me thinking too much .... I want the world to know about these issues. The campaign to release me and other activists was so important, not only because it gave us strength but because it shone a light on the injustices we were fighting against. The Cambodian authorities would prefer these injustices to happen in the dark .... I am really proud of what I have been able to do for the Boeung Kak community. I was just an ordinary housewife, an ordinary person with a limited education and limited knowledge of social issues and the world. We got solutions for almost everyone. This is what I am proud of, even though I have been jailed .... When we face problems and we lose hope, we become weak and can give up what we are struggling for. So, I want people to not feel hopeless (Tep Vanny cited in Amnesty International 2018).

As I have written elsewhere, it is my belief that BKL women's actions have given rise to a less fatalistic, albeit dangerous, era in which human rights abuses do not go unwatched (Brickell 2014a). They have ensured that the Cambodian government sits in an incongruous and uncomfortable moral position of being seen to punish wives and mothers for their commitment to the survival of their households. BKL women have pioneered new ways in Cambodia for women to refuse the ordinary of violence. Their actions rendered vulnerable the government's goal to keep injustices in the dark, and consolidate control in the everyday lifeworlds of its citizens.

Conclusion

Together domestic violence and forced eviction show how the violent enactment of law and lawfare can be clearly witnessed in Cambodia. Positioning the omission and commission of law in 'a field of pain and death' (Cover 1986, p. 1601), the chapter has demonstrated the wide-ranging consequences that its weaponisation has had in different women's lives. Its significance in the intimate wars faced by women and the unrelenting crisis ordinary they manage has been demonstrated in three main ways.

First, the chapter has examined the amplification of violence through a hybrid attention to the omissions and commissions of law. In respect to domestic violence, the Cambodian government has not taken effective steps to enable women to access their claim rights to a life free from violence, thus committing acts that I referred to as 'commissioned omission'. As Joel Brinkley (2011, pp. 233 and 234) writes of DV law in Cambodia, 'Like so many other laws the rulers did not like, the government simply declined to enforce it .... The government had been shamed into passing a law it did not agree with. Its solution, as usual, was to do nothing.' Even if the government is not killing women directly, this approach promotes conditions of impunity that increase risks of victimisation by normalising violence against women. The reliance on local reconciliation, and its potential for administrative violence, is also a commissioned and legally co-opted lever to keep couples together. But for many women this modus operandi represents an ambiguous departure from intimate war. Women's unfinished, often never-ending, journeys to lives free from domestic violence, are experiential processes that support my argument for viewing Cambodian society through the prism of the crisis ordinary. This crisis ordinary is typically carried on the soles of women's feet, pain embodied and hidden, away from public view. As artwork (Figure 6.10) produced to mark the '16 Days of Activism against Gender Violence' reads: 'One powerful story was about a husband who worked very hard, but didn't understand how busy his wife was at home, watching the kid, cleaning the house, and preparing everything. He grew angry and slowly started to argue with, and eventually fight his wife'.[24]

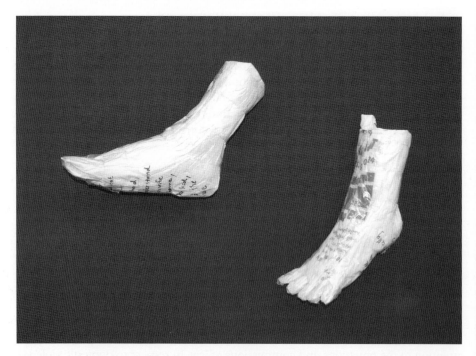

Figure 6.10 'Footprints' artwork by Cambodian schoolchildren under the direction of Heak Pheary, 2016. Source: Courtesy of NGO-CEDAW.

In the case of forced eviction, overlapping acts of omission and commission are multiply lived and presenced in more overt ways than domestic violence. From their hands held up high, to the blood streaming down their faces, BKL women have embodied the injustices and violences they face in public view.

They have been excluded from official reporting mechanisms, and actions on violence against women given the violence they have encountered is, in part, state-sponsored and thus highly politically sensitive. Lawfare has been intentionally used as a means to achieve other violences of retribution including those of incarceration and solidarity-breaking. Their experiences over the past ten years, Amnesty International (2017, p. 5) writes, 'provide an important illustrative example [on the latter front] of how the criminal justice system has been used in an apparent attempt to dismantle a tightly knit nucleus of activists and to limit their activism'. As geographers Rachel Pain and Lynn Staeheli (2014, p. 345) poignantly suggest, 'resistance may undo violence and create further forms at the same time'. BKL women's calls for change – their bringing of 'fire' from the private into the public domain – was considered too threatening to existing distributions of material and symbolic power by the Cambodian government. Their publicly articulated resistance has been met with the episodic denial of their freedom. Law 'has the power to pull individuals to particular territories, to make them disappear from others, and even to place them outside of nations altogether, thus exiling them from "homes" in multiple senses' (Coutin 2016, p. 7). In the BKL case, imprisoned women have experienced multiple displacement over time and space, from possible eviction, taking to the streets away from their homes and family members, and enduring wholescale isolation from them during periods of incarceration. The violences they have encountered 'is to terrorize, to send a message, first to women and then, through them, to entire populations that no mercy should be expected' (Federici 2018b, p. 50) when challenges to the current order are made.

Second, this chapter has developed the argument that the crisis ordinary is an ontological fissure that victims of domestic violence and forced eviction navigate in their intimate lives. These tears are far from temporal abnormalities but rather are hewn with the spatiality and temporalities of women's lives. The durative work of upholding the pretence of family harmony is part and parcel of the crisis ordinary lived by domestic violence survivors who have few alternative options than to stay with abusive partners. While the stack of promissory notes is a material cue to the crisis ordinary contained to the home through administrative means, BKL women's contesting of forced eviction led to domestic life being lived on the street. The bed frame parked across lanes of traffic speaks to the range of disobedient objects that women brought from their threatened homes to occupy Phnom Penh streetscapes and disrupt its everyday flow. Their activism became a new ordinary which the Cambodian government grew increasingly fearful of, concerned that it would galvanise a wider populace to ask if a more courageous and just everyday was possible. Their 'wakefulness' (Sharpe 2016, p. 4) had the potential to spark a consciousness too subversive of elite interests. The radical potential of the mundane bed frame and its decentring from the domestic home to inhabit and command the street is also demonstrative of the selective illegalisation of everyday life for political ends. Its placement and rendering illegal of the previously accepted – the country's unenforced Traffic Law – transformed impotency into lawfare against BKL activists and their rights to dwell in public and private space. While BKL women foisted a crisis ordinary on the Cambodian government through their regular, sometimes daily, activism and this had some positive impacts (see Chapter 5), over time its weight took its toll on group members and their relationships with one another. Dwelling in a protracted state of the crisis ordinary can leave people 'worn out' and ultimately have a depoliticising effect in undercutting the potential for macrolevel change (Aslam 2017, p. 8).

Third, taking into account these prior points, the chapter reignited discussion of citizenship and its gendered contingency. BKL women's jailing reflects the point made by Chloé Buire and Lynn Staeheli (2017, p. 137) that 'such acts [of active citizenship] are required to draw attention to a cause, but they might violate norms of civility ... and might not be consistent with the kinds of citizenship that are promoted through the pedagogy'. BKL women have challenged the pedagogy of the Cambodian government intelligible within hierarchical norms of 'keeping to one's station' and thus the status quo of their domination. With this, women's acts of citizenship are performed under precarious circumstances in which their exercising of rights carries a series of gender-differentiated risks and rejections. In contrast to the lack of legal sanction a perpetrator of domestic violence is likely to encounter, victims often find themselves unable to seek justice due to risk factors such as social stigma, family dishonour, and an indifferent reaction from legal authorities. They are typically unable to access justice and are stuck with a sense of entrapment. BKL women meanwhile are hauled quickly into court and jailed on account of their peaceful activism.

As such, the *Phnom Penh Post* headline, 'In Cambodia's courts, it's a man's world' rings true (Kijewski and Sineat 2018). The pain of the legal order and persistent impunity in Cambodia is thus embodied within and on the wounded bodies of its female citizenry and the enflamed homes that are the spatial crux of the crisis ordinary.

Endnotes

1 Positive peace is 'a social condition in which exploitation is minimised or eliminated and in which there is neither overt violence nor the more subtle phenomenon of underlying structural violence' (Barash and Webel 2009, p. 7).

2 This includes basic necessities, including petrol provided to travel locally via motorbike and respond (see Brickell 2017b and Brickell, Prak, and Poch 2014 for more information on the inadequacies of state resource provision).

3 When intimate partners promise to change, this is a 'charm syndrome', which is used to sustain women's hope but ultimately keeps them in a cycle of violence (Radford and Hester 2006, p. 4).

4 As LICADHO (2017) found in their review of 400 cases they investigated between 2014 and 2016, 'LICADHO staff have seen many cases in which couples have made

multiple contracts with the police. There is no legal basis for such contracts and the Criminal Procedure Code explicitly prohibits police from not acting on a case where the victim has withdrawn the complaint or there has been a negotiated settlement between the victim and the suspect. Despite this, the practice of reconciliation by the police is routine' (p. 5).

5 Johan Galtung (2001) outlines four approaches for peace workers to adopt. He caveats in the conclusion, however, that no one approach is a panacea, that none can singularly close cycles of violence and thus advocates for a combination.

6 In her work on domestic violence and indigenous jurisprudence, Donna Coker (2006, p. 1362) argues, for contrast, that restorative justice 'may be beneficial for some women who experience domestic violence, but only if those processes meet five criteria: prioritize victim safety over batterer rehabilitation; offer material as well as social supports for victims; work as part of a coordinated community response; engage normative judgments that oppose gendered domination as well as violence; and do not make forgiveness a goal of the process'.

7 Yet as the Explanatory Notes (RGC 2007c, p. 196) concede, the terms 'minor' and 'severe' are not defined in either and are therefore open to interpretation. It is left to these supplementary guidelines to set out the cursory idea that, 'Repeated commission of domestic violence, special vulnerability of the victim, combined domestic violence such as physical and economic violence, or the involvement of more than one perpetrator or victim should be regarded as circumstances that leads to characterizing any misdemeanour as "severe"' (RGC 2007c, p. 197).

8 Her already precarious position was also not helped by her hypoglycaemia and heart problems exacerbated by manual labour collecting aluminium cans for recycling.

9 According to the United Nations Office on Drugs and Crime (UNODC) (2013, p. 137) 2010 statistics, 55% of female homicide victims in Asia are killed by their family members or intimate partners; the figure for men, on the other hand, is only 6%. While the study does not provide figures for Cambodia on murder by an intimate partner specifically, it does include 2009 data, which shows that intentional homicide victims are 75.3% male and 24.4% female in the country (UNODC 2013).

10 Husbands have also been injured in violent clashes, as have monks, by security personnel.

11 The use of electrified riot shields and water cannons by authorities and in some cases private security personnel employed by the Municipality of Phnom Penh for use in confrontations and detentions of activists was raised by the UN Special Rapporteur for Human Rights (United Nations General Assembly 2013, p. 15).

12 The Fourth World Conference on Women: Action for Equality, Development and Peace was the name given for a conference convened by the United Nations during 4–15 September 1995 in Beijing, China.

13 In January 2014, it was reported that five garment workers died at a strike protest from open gunfire by the Cambodian military police using AK-47s (Soenthrith 2014).

14 For more information see: http://www.un.org/en/events/endviolenceday.

15 According to the acknowledgments at the end of the report, The Cambodia Gender Assessment (Ministry of Women's Affairs (MOWA) 2014b, p. 28) 'was produced under the overall leadership and coordination of the Ministry of Women's Affairs, with support and contributions from Government Line Ministries, Development Partners, and Civil Society'.

16 In their study of 600 women, The Cambodian Center for Human Rights (CCHR) (2016) found that 51.1% of the women claimed that land conflicts had impacted upon their familial relationships.

17 The survey undertaken by the Cambodian Center for Human Rights (CCHR) (2016) echoes this sentiment. It found that 23% of women in relationships that were involved in land conflict self-identified as victims of domestic abuse and of these 53.9% had never experienced this violence prior to the land conflict. See Beban and Pou (2014) for additional discussion of these connections.

18 '13 Boeung Kak activists violently arrested after breaking into song,' video report, 22 May 2012, http://www.licadho-cambodia.org/video.php?perm=31.

19 https://www.licadho-cambodia.org/flashnews.php?perm=184.

20 In respect to their emboldening, the headline of *The Cambodia Daily* newspaper article (Narim 2012) reads 'Released Boeung Kak Women Back to Old Ways' and comments upon their renewed and unrelenting protest only two weeks on from the hard-fought release of the 'BKL 13' from prison.

21 See also Raka Shome (2011) for a critical perspective on the representational logics of whiteness in which contemporary white women, such as Clinton, are positioned as 'global' mothers in popular culture.

22 In 2004 Chea Vichea, a prominent trade union leader, was shot dead as he browsed a newspaper stall on a busy city street. In 2012 Chhut Wutty, a prominent environmental activist, was killed at a police checkpoint while aiding two journalists in their witnessing of deforestation in the Cardamom mountain forests (see the film *I Am Chut Wutty* for more information). In 2016 Kem Ley, a political commentator, was murdered at a petrol station in Phnom Penh. Kem Ley's wife, Bou Rachana, has since been granted asylum in Australia with their five sons. She has received a death threat while living there (Handley 2018).

23 For more information see: http://ngocedaw.org/dignity-project.

Chapter Seven
Dwelling in the Crisis Ordinary

Introduction

That the absence of war does not mean peace (Galtung 1969) holds particular salience for Cambodian women and their families. While Cambodia is perhaps best known for the familial incursions of the Khmer Rouge genocide (1975–1979), *Home SOS* has rendered clear how intimate war has never really ceased. Although the fields of conflict may have been reshaped and rethought through new types and logics of warfare, the home has remained a primary warzone. The recesses of home, and the bodies that dwell in it, have continued to be subject to 'intricate invasions' (Bhabha 1994, p. 13). The book has thereby illustrated the value of exploring the (un)eventful time-spaces and performances that are outside, but nonetheless connected to, formal war. It has positioned the home as the aperture through which to trace and critically discuss the embodied experiences of women along a continuum and combination of violence and peace in neoliberal times.

Silvia Federici's (2018b, p. 46) argument that 'the evidence is mounting that a new war is being waged against women' has thus been interrogated in the perennially under-studied context of Cambodia, where this war can be observed in, and through, the extra-domestic home. BKL women's experiences have perhaps most directly spoken to the wider trend she identifies, namely '*the new violence against women is rooted in structural trends that are constitutive of capitalist development and state power in all times*' (Federici 2018b, p. 47, emphasis in original). It does so,

Home SOS: Gender, Violence, and Survival in Crisis Ordinary Cambodia, First Edition. Katherine Brickell.
© 2020 Royal Geographical Society (with the Institute of British Geographers). Published 2020 by John Wiley & Sons Ltd.

Federici contends, 'by attacking people's means of reproduction and instituting a regime of permanent warfare' (pp. 49–50). For the developing capitalist order of Cambodia to be unchallenged, intimate life, with women at its heart, is subjected to bio-necropolitical impulses and pressures that cannot be ignored. Even though Cambodia's recent history is so often told through its 'battlefield to marketplace' transition (see Chapter 3), *Home SOS* has shown how women's experiences of domestic life problematise this motif of change.

The home is the most basic, yet vital, space of reproduction and, as such, the responsibilities and relationships that sustain it have been shown to be targeted for exploitation. Gendered forms of precarity and slow violence, which call on women to hold their ground under multiple wars of attrition, have solidified in multiple realms of home life, from domestic violence to forced eviction. Having informed and stimulated geographical scholarship on imperilled and pried apart homes, the book has been attentive to how women maintain their binding to modes of domestic life under threat. The bubbling of crisis ordinariness beneath, and up from, the surface has shown how women are considered, and often consider themselves, responsible for the 'reproduction of predictable life' (Berlant 2007, p. 754). The book has therefore focused on the spectre of ordinary and elephantine violence (Brickell and Maddrell 2016; Tyner 2016a) still too often forgotten or relegated to the side lines of political geography and geographies of violence scholarship.

'Positive peace' requires that structural issues which give rise to such violence are proactively addressed, yet over the course of writing *Home SOS*, the prospect for this has diminished as Cambodia has further embraced capitalism and taken a backward slide into authoritarianism. Starting from the domestic sphere, the book has provided an alternative reading of Cambodia told through the creative survival-work that is being undertaken against the backdrop of domestic violence, forced eviction, and this evolving political economic landscape. As a result, the book has told new spatial histories of Cambodia through the stories of Khmer women and the precarity of their homes under diverse yet interconnected circumstances. The empirical material has also provided glimpses of crisis ordinaries that other vulnerable groups are experiencing, including evicted men, whose experiences warrant fuller telling. While literature has tended to focus on women and their children in the study of displacement, there are mounting calls for men and boys to be the focus of more sustained study (Brun 2000; Horn and Parekh 2018; Suerbaum 2018). This is certainly an area for future research, which I endorse and hope to build on myself moving forward.

For now, however, women's encounters with marriage and its breakdown have been shown as an important proxy for seeing fine-grained workings of a nation and its intimate governance. *Home SOS* has illuminated how confinements to, and displacements from, marriage are lived and felt concurrently through the indispensable forcing back together of marriages hewn with domestic violence, and those of forced eviction activists deemed dispensable and prized apart.

However, while these represent divergent political contingencies and spatialities of marriage, taken together they embody stringent forms of social control targeted at women and pursued to contain and displace any contest to capital and patriarchal control. The entanglement of these structures and the attempted denial of women's autonomy through them runs across the domestic violence and forced eviction cases presented. Their crisis ordinary bonds wartime and peacetime violence in Cambodian women's lives such that 'violent conditions [continue to] burn in the background of daily life' (Lawrie and Shaw 2018, p. 8).

Home SOS has thus risen to the challenge 'to recognise the interconnectedness of forms of violence that we do not always recognise as being connected or, for that matter, as being forms of violence' (Sassen no date). Just as James Tyner (2019, p. 1306) has recently argued that 'it is necessary to reconfigure gendered and sexual violence not as apart from, or autonomous to, the broader political economy of Democratic Kampuchea but rather as an integral and systemic relation of social reproduction', it is imperative that the gender-based violences of today are also understood to be integral, rather than superfluous to, the operation and hegemony of crisis ordinary Cambodia. Direct and structural violences are mutually constitutive of the intimate wars being mobilised against women on several fronts. *Home SOS* has thereby contributed to critical geographical engagement in human rights grounded in the dissonance established between their idealism and shallow roots. It has done so in a vernacular heavily determined by the neoliberal and patriarchal political economy. It showed that while human rights are inadequate provisions for enabling women's homes and bodies to be free from violence, they are nonetheless important forms of rights-talk and humanising influences in Cambodia. 'Increasing claims that the time of this discourse as an emancipatory tool is up' (McNeilly 2018, p. 1) are therefore tempered by women's more ambivalent experiences at grassroots level.

In the concluding discussion that follows I bring together and further detail the key findings of the book. Three sections are provided, the first on thinking through the crisis ordinary ('inhabiting the fire'); the second on political geographies of home ('fanning the fire'); and the third on working towards a feminist legal geographies project to develop future work ('legislating the fire'). The chapter therefore goes full circle to the book's beginning to evoke the multiple practices of firefighting taking place in contemporary Cambodian society.

Inhabiting the Fire: Dwelling in the Crisis Ordinary

A core argument of *Home SOS* has been that domestic violence and forced eviction are manifestations of the crisis ordinary, a condition that characterises home life for many Cambodians, but especially for women. To this end, the book has revealed how domestic life is not a matter of 'carrying on as usual' amongst the violences that women encounter, but is actually the incorporation of disruption into 'business as usual' (Calvente and Smicker 2017, p. 6). Writing on Black Lives Matter, Lisa Calvente and Josh Smicker (2017) argue that the movement's success in exposing everyday realities of racialised violence resulted in the ritualisation of protests and actions aimed at disrupting daily life. That the repetition of crises become normalised disruptions has also been detailed in the preceding chapters of *Home SOS* focused on deepening scholarly understanding of intimate geographies of violence.

Akin to intimate war, which highlights the imbalance between 'high profile, large scale, international armed conflict over small scale, domestic and "personal" insecurity' (Little 2019, p. 2), the lens of the crisis ordinary has brought to the fore the contingency of any stability and safety that the home presupposes to offer. The survival-work that women undertake in the home aims to meet, but also to exceed baseline expectations of physical shelter, to provide a domestic environment that sustains the 'good life'. As Susan Fraiman (2017, p. 166) argues, reading about domestic struggles reboots respect for domesticity: 'the need to secure ourselves and our things; the impulse to beautify; the comfort of going about daily rounds; the importance of warm, tactile connections; the sense of belonging and obligation'. Together domestic violence and forced eviction demonstrate how these attributes are far from guaranteed; rights to dwell free from violence remain elusive for too many and are to be fought for.

In Cambodia, the crisis ordinary has become a violently predictable and constitutive aspect of domestic harmony performed at the edge of life and death. 'Housing is a human rights issue – it makes or breaks us. It is the difference between life and death', Leilani Farha (2018, np), UN Special Rapporteur for Housing, spells out. This truism is born out in *Home SOS* but is also complicated. As Chapter 4 demonstrated, domestic violence and forced eviction are intimate wars that Cambodian women are traversing through encounters with physical, social, and figurative forms of death. Their survival-work and the dependency of households on it, is part of a 'feminization of survival' (Sassen 2000, p. 506) usually linked to women's international labour migration. It is also evocative of the 'parasitic dependence on the free appropriation of nature and the body and work of home' (Federici 2014, p. x), which is exercised under capitalism as the 'avatar of patriarchy' (Mies 2014, p. xxiv).

That resilience in this milieu has become necessary to the workings of domestic life, to very liveability, is deeply problematic. Responsibilisation, subjectification, and naturalised uncertainty are all burdens that women disproportionately share in neoliberal Cambodia. Lauren Berlant (2018, np) observes, in this regard, that 'living as the negative subject of any inequality requires so much creative energy to be taken up in microadjustments and improvised defenses'. Cambodian women are charged with being resilient subjects with an elasticity and patience to manage 'fire in the house'. Many respondents have become hyper-vigilant to the ongoing survival-work needed and this, in turn, becomes a mode of being in the world that reproduces itself.

The empirical stories told in *Home SOS* therefore raise challenging questions for theoretical agendas linked to the normative ascendency of resilience in scholarly and political discourses. Namely, the material presented in the book points to the need for critical engagement with the idea of resilience and the politics it stirs. It has made plain the realisation that 'violence is first of all authoritarian. It begins with the premise: I have the right to control you' (Solnett 2013, np). Cultural and economic logics of governance in Cambodia hold women in the intimate embrace of the crisis ordinary in time and space under the onerous weight of spousal violence and forced eviction. If resilience is taken as 'the capacity of a system to absorb disturbance' (Walker and Salt 2006, p. 62), then this thinking has the potential to close off, rather than open up, wider questions of structural violence, rights to dwell, and strives for progressive social change in this sphere. What the crisis ordinary does, that resilience struggles to, is to keep crisis on the discursive table rather than sweeping it underneath. The crisis ordinary re-questions and re-positions mundane registers of everyday life as domestic emergencies that are sustained embodiments of failure. It encapsulates the infusion and normalisation of precarity as a condition, dynamic, and affective labour of domestic life. That resilience conceives of human life so that it adapts, rather than resists, to the conditions of suffering in the world (MacKinnon and Derickson 2012; Reid 2012; Welsh 2014) means that it risks sustaining rather than challenging violence. The crisis ordinary does not exclude resistance, rather it points to its own potential co-option, into the survival-work of the everyday. Given the intensification of homemaking required by conditions of precarity, studying domestic violence and forced eviction in confluence has evidenced, particularly strongly, this co-option of survival-work into the ordinary.

As a consequence, *Home SOS* has evoked the conceptual utility of slow violence in drawing attention to the gradual unfolding, rather than (only) immediate unleashing, of violence and survival-work in women's lives. More broadly, I am not alone in sensing this possibility in geography (Anderson et al. 2019; Davies 2019; Pain 2019). Thom Davies (2019, p. 4) concludes, for instance, that slow violence has 'much to offer as a temporal and spatial concept' and therefore 'the door is surely ajar for more sustained engagement with slow violence from a geographical perspective'. Emerging work on the slow explores peoples' 'striving to survive in and through risky, inhospitable landscapes' (Anderson et al. 2019, p. 7), with my own research having taken a domestic turn to examine the attritional and depletive impacts of domestic violence and forced eviction over the short and longer term. As such I have been wary of the usual focus on 'temporal urgency and newness', which 'subtends the slower and less visible but everyday' (Adey forthcoming, np). The recently published scholastic endeavours cited here pick up from Rob Nixon's (2011) seminal writing on slow violence and alerts researchers to the accumulating damage of violence, which can appear too mundane, or hidden in plain sight, to be registered. But it must.

In this respect, *Home SOS* has propelled work in geography on routine and/or durative crises from a gendered perspective. If 'gender is a kind of doing, an incessant activity performed ... a practice of improvisation within a scene of constraint' (Butler 2004, p. 1), then so too is the survival-work of dwelling in this crisis ordinary. In the majority of cases presented, the book has shown how women's means of survival are rooted in a concern for upholding gender norms that convey recognition and respect under adverse and dignity-eroding circumstances. While focus in the development sector tends to be on how discriminatory norms combine with other forces to limit women's choices and well-being – to realise their full potential and avoid harm – the absence of structural commitment to facilitate or support these changes means that norms both permit, as well as restrict, the conditions of life itself. In other words, norms can support, as well as undermine, the capacity to persevere in the crisis ordinary. For this reason, women switch between gender scripts to mitigate violence and death the best they can. Any critique of gender norms must, henceforth, 'be situated within the context of lives as they are lived and must be guided by the question of what maximizes the possibilities for a liveable life, what minimizes the possibility of unbearable life, or indeed, social or literal death' (Butler 2004, p. 8).

Victims' help-seeking, be this connected to domestic violence and/or forced eviction, needs to be viewed as a political tussle in which hegemonic power relations and the spatial norms of privacy are transgressed. Reporting spousal abuse at commune level, for example, might be a 'small' or 'quiet' act, but it can 'still contribute to a wider process that ultimately may lead to change' (Pain 2014, p. 143) given that silence and stoicism remain expected of Cambodian women despite legal changes. Furthermore, while women's acts of resilience, of 'getting by' in the crisis ordinary, might indicate inertia or toleration, given the bio-necropolitical dynamics presented through the book, their performance is a production that can never be guaranteed (De Verteuil and Golubchikov 2016).

The book has brought a range of persons and objects in crisis ordinary settings into contention to explore these performances further. The materiality of home has emerged not only as an important means for improving understanding of precarity as an experience but also as a possible rallying point for resistance with political potential (Waite 2009). Alternating between domestic violence and forced eviction examples, these speak variously to the coercive and agentic potency of things. They range from: the weight of a metal torch brought down on a woman's head, to a nest full of eggs balanced on an activist's head; a house cut in half to avoid legal wrangling over divorce, to the shattered brick line of a BKL house once sheltering a family; the pink balloons of a banished campaign cycle ride, to those released into the sky in an eviction protest; a wicker basket of plates rattling as marital tensions are normalised and contained, to a bed frame dragged out of place on to a busy street; the batter of women's rights spilling out of the cake pan, to the tip of the lotus flower thrust upwards in peaceful defiance.

The matter of survival is not passive; it offers a lively, audacious, and even dark (re)telling of political geographies of home and (dis)harmony rooted in the stories of objects and their proximation to the intimate wars of domestic violence and forced eviction. The extra-domestic is site to violent and disobedient objects used to exert hegemonic power and its creative countenance. Geographies of violence literatures and the technological infrastructures of formal war, usually told from above, thus require a regrounding in the banal materialities of intimate wars also taking place at official peace below. To do this, the book points towards rich seams of interdisciplinary dialogue that could be forged between scholarship emerging from geographies of violence, feminist geopolitics, and (feminist) new materialism.

Through its integrated study of domestic violence and forced eviction, *Home SOS* has demonstrated therefore the experiential, performative, and material corollaries of domestic life that are characteristic of contemporary Cambodia. I have grounded Lauren Berlant's conceptual work on the crisis ordinary within a gendered rewriting of intimate lives played out across violence and peace. The home, I have argued, is an impasse shaped by crises that Cambodian women perform as an everyday practice and means to live on in ever-challenging times.

Fanning the Fire: Political Geographies of Home

Home SOS has fleshed out how politics is influenced by, and emerges from, the home (Brickell 2012a) as a space indelibly connected to conflicts over power, resources, autonomy, and agency (Madden and Marcuse 2016). Following the logic of the Cambodian government, ordinary life has to dwell within an affective landscape and management of fear to uphold their dominance. To achieve this, human rights are positioned as a source *of crisis*, not as infringements and sufferings, that result from government (in)actions. My observations have connected to other political analyses of Cambodia and its growing reliance 'on the idea of national security as a strategy to justify its power consolidation' (Peou 2018, p. 81). Home, family, and nation operate in the same 'mythic metaphorical field' (Bammer 1992, p. x) in which the Cambodian state is set against a corrosive outside that threatens absolute sovereign power. For the Cambodian government at least, violence or 'fire' in Cambodian homes is none of their internal doing but is externally fuelled. This blame game tries to divert attention away from the bio-necropolitical manoeuvres of its own governmental organs.

That 'uncivil' acts are painted as leading to all-out crisis, are governmental reminders of the necessity for the ordinary of subordination. The choice, cum threat, set out is a stark one – inhabit the crisis ordinary or risk a return to formal war. If Cambodian women are cast as the guarantors of harmony, then the ruling party position themselves as the guarantors of stability (Chapter 3 discussed this trope of order and stability). Under these circumstances, however, Jin Haritaworn, Kuntsman, and Posocco (2016, p. 2) contend that the 'distinction between war and peace dissolves in the face of the banality of death in "zones of abandonment"'. The crisis ordinary is a compensatory practice and state that mediates between home as the site of life and death. It is an overwhelming ordinary that brings to the fore the intertwined bio-necropolitics of domestic and marital life.

In Chapter 5 I detailed how the Cambodian government has been unwilling to sanction women's contestation of domestic violence or forced eviction that might challenge prevailing hegemonic power relations. Given the scale of domestic violence in Cambodia, and despite its 2005 law, the state has placed the onus on victims' adaptive capacity in dealing with the undergirding violence and vulnerability that women face within their homes. Their domestic adversity and the reproductive labour necessitated by it is called upon by elite men to ensure the harmony of the wider nation, and, by extension, the sacrosanctity of their power. However, the ruling party has had a problem in the latter case of forced eviction. BKL women became a public metaphor for Cambodia's larger, perennial problems, injustices, and tragedies. Unlike domestic violence victims and their advocates who have, except for dedicated days and backstage bargaining, remained largely out of sight, BKL women have become highly visible actors on the national and international stage. They have forged 'new cartographies of resistance, struggle and survival' (Bailey and Shabazz 2014, p. 449) in publics deemed too proximate to international bases of power. Their closeness to suffering and violence on a daily and eventful basis and their defiance has also greatly concerned the government who have been increasingly dealing with their own crisis of legitimacy. The front cover of *Home SOS* evokes women's public pushing back against the muscle of the state; the violet painted nails of an elderly BKL activist, held firm against the riot shield trying to repel her. BKL women's novel and affective 'politics in the first person' approach (Mies 2014, p. 28) has personalised power for counter-hegemonic purposes.

The 2013 national election had brought the ordinary of poverty, housing loss, and endemic corruption to the fore in a closely run vote. The CPP needed to put out the fires that BKL women were regularly igniting, to avoid being burned by a wider turn to revolt. Authorities therefore increasingly misused the justice system to silence human rights defenders and critical voices in the lead up to the next election in 2018. State security has been predicated on the circuitry of moving some subjects into modes of security and others into abandonment in order to pursue its goals (Rose 2000). The temporal contingency of these movements has been revealed, as has their significance, for the denial and then authorisation of women's liberty at apposite times for political gain. Human rights are regulative ideals that the Cambodian government has been forced to negotiate as part of its time-sensitive national and international political manoeuvrings. Women's homes and bodies are caught up in the times and rhythms of these high-order politics, not separated from them.

Indeed, the gendered dynamics of the crisis ordinary and their connections to these high-order politics have been multiply shown in *Home SOS*. In this respect, the mounting significance of debt in the narratives of both domestic violence and forced eviction victims revealed in the preceding chapters are telling. If 'resilience assumes a future of inevitable and worsening crisis and seeks only to minimize its effects, adapting to changing conditions so as to keep existing socioeconomic conditions of liberal life the same (or perhaps more accurately, on life support)' (Wakefield 2018, p. 6), then the taking on of household debt is a key mechanism for propping up the social reproduction of Cambodian households and their 'resilience'. A slow crisis hidden in plain sight has been building for some time in Cambodia. Borrowers are increasingly experiencing problems of over-indebtedness, compelling families to reduce food consumption, take out new loans to service prior debts, migrate, and sell their land in distress (Green and Bylander forthcoming). It is not only that 'women are traditionally charged with financial management within Cambodian household' and 'the strains created by over-indebtedness are often disproportionately borne by women', but also that 'MFIs disproportionately lend to women, with the largest MFIs all reporting that the majority of their borrowers (in some cases the vast majority) are women' (Green and Bylander forthcoming, np).

The precariousness of Cambodia's recent economic success comes into sharp relief when the impacts of this most local of neoliberal policy interventions are examined (Bateman et al. 2019). That a recently published report on farmers selling land to repay microfinance loans (Sahmakum Teang Tnaut and LICADHO 2019) received such force of government and industry backlash (Narim 2019) speaks to political concerns and sensitivities about the possibility for crisis this trend has.[1] *Home SOS* has begun to illustrate how debt has an intimate weave in marital and domestic crises unfolding in situations of both domestic violence and forced eviction. Given the profiteering logics of MFIs, the gendered profile of borrowers, and the onus of survival-work on women in Cambodian society, it is important that future work opens up conventional discussions of gender-based violence to question and expand the narrow parameters of what is currently understood as economic violence. Geographers have a role to play in researching the 'more-than-economic geography' of debt, debt spaces (Harker 2017, p. 600) and violence, with the extra-domestic home demonstrative of this potential. This is an area of study that is gathering momentum and will continue to do so given the accelerating financialisation of home life (García-Lamarca and Kaika 2016; Forrest 2015) and poverty (Mader 2015; Roy 2010) being witnessed globally.

Cambodian women, for this and a whole host of other reasons, intensely feel and manage 'the precarities of the ongoing disaster[s]' (Sharpe 2016, p. 5). I have argued that while domestic violence is widely understood as gender-based violence, forced eviction should also be understood as such. A key contribution of the book, therefore, has been to further explode the range of what is considered and counted as gender-based violence. As Sally Engle Merry (2009, p. 3) noted over a decade ago, 'International activists continue to expand the scope of violence against women, to include cultural practices such as female genital cutting, illegal acts such as dowry deaths, the trafficking of women as sex workers, [and] the effects of internal wars such as displaced people.' Listening to women contesting forced eviction in Cambodia means being attentive to their calls for the violence they experience as human rights defenders to be counted, and responded to, as gender-based violence. Who falls officially in, and out, of this categorisation matters at a time of unprecedented numbers of women being injured or killed for their human rights work across the world. BKL women's calls, for example, are part of a wider push for women human rights defenders to be protected rather than attacked (Office of the High Commissioner for Human Rights (OHCHR) 2018, 2019). In November 2018 on International Women Human Rights Defenders Day, the Office of the United Nations High Commissioner for Human Rights (OHCHR 2018, np) published a press release stating their condemnation of the violences they face worldwide:

> The current global context of unchecked authoritarianism as well as the rise of populism, of corporate power and of fundamentalist groups are contributing towards closing the space for civil society. This is being done through the enactment of laws and practices that effectively impede human rights work, including the misapplication of certain laws such as counter-terrorism and public assembly laws. In this context, women human rights defenders face additional barriers of economic and structural discrimination and unique challenges driven by deep-rooted discrimination against women and stereotypes entrenched in patriarchal societies related to gender and sexuality.

Three months after these concerns were shared in the public domain, a new report published again by OHCHR (2019) went so far as to say that violence against women human rights defenders by actors on behalf of states had become 'normalised'. Chapter 6 bore these concerns out most closely in relation to BKL women's activism, given the habitual violence they encountered as targets of state aggression and their related and repeated imprisonment on such spurious grounds. BKL women have been exposed to what is tantamount to gender-based violence, not just because of the direct violence they have faced from state forces and in some cases, their husbands, but because home loss is also a challenge to roles and identities that women interviewees value. The pre-occupation of political economy approaches to eviction that privilege 'macro narratives' and broader processes of capitalist accumulation by dispossession, thus risk side lining the 'micro-politics' that are also at play (Lancione 2017, p. 1015). Moreover, given growing evidence that 'brutalizing women is functional to the "new enclosures"' (Federici 2018b, p. 50), it is imperative that macro-level analyses attend not only to capitalism but also patriarchy. As Chapter 1 ventured, it is perhaps of little surprise that women's pushback against forced eviction has offended both.

As a result, one of my goals has been to write about the violence that BKL women experienced as gender-based violence and to give some parity to the level of institutional attention given to domestic violence. Far ahead of the 2017 viral spread of the #MeToo movement against sexual harassment and assault, BKL had forged their own #MeToo-style moment. They had realised the violence they encountered in their protest was not to be recognised as such by a system of national laws and human rights norms, which offered few protections against a political economy that selectively emphasised its concern for sanctioned forms of gender-based violence. This 'hierarchy of harms' (Kirby 2015, p. 463) within gender-based violence, is something that *Home SOS* has provocatively challenged.

Legislating the Fire: Towards a Feminist Legal Geographies Project

In the introduction to *Home SOS*, I identified law as a thread that I would run through the book. Chapter 6 focused most explicitly on the intimate wounds of law and lawfare being enacted against women, but the legal paths and broader struggles are raised throughout. With regard to domestic violence law (DV law), for example, in Chapter 5 I suggested that its ratification for the upper echelons of the Cambodian government was never about the building of a new legal infrastructure capable of transformative change for women. As Berlant (2016, p. 393) explains, this fixing of broken infrastructure – and thus the absence of DV law – can actually generate 'a form from within brokenness'. In other words, DV law and its truncated penalties derived from the same lifeworld of structural violence as before. Legal rights are meaningful only if they can be asserted and an essential component of the rule of law is that women can actively claim the entire range of rights. 'Legal interpretation must [be] capable of transforming itself into action' as Robert Cover (1986, p. 1617) has famously written. Given the importance of tracing how legal norms are forged, interpreted, and applied, the book has shown how critical it is that the animated legal journeys, and not just outcomes, of law are the focus of scholarly work. By combining domestic violence and forced eviction in one book, the masquerades of law and lawfare have been revealed.

DV law in Cambodia has been shown to be a tactic that rearranges just enough to maintain the current arrangement of the crisis ordinary. It has been a hollow narrative performance of acceptance to international norms that has adoptive risks for women given the likelihood of stigma, shame, corruption, and other financial precarities. Focusing attention on lived experience helps to go beyond naturalised assumptions about law to provide a deeper understanding of what takes place in women's encounters with its pluri-legal manifestations at home and in wider society. For many women, a 'justice facade' exists – what can be described as 'a distanced and surface-level view that masks power and the complexities of everyday experience' (Hinton 2018, p. 7). By combining the study of domestic violence and forced eviction, the book has been able to venture behind this facade in twin realms to reveal the penalising journeys that women are either precluded from taking in the first place, or suffer, in seeking legal action. Questioning how meaningful law is, Federici (2018b, p. 56) notes that there is a growing sense that demands for more punitive legislation for violence against women globally is a 'dead-end'; it 'only serves to give more power to the very authorities that are directly or indirectly responsible for the problem'. More effective, she argues, are 'the strategies that women devise when they take things in their hands' (p. 57). Yet taking things into their own hands, BKL women who fought against forced eviction found themselves (as I described earlier) on the receiving end of judicial harassment and were incarcerated for displaying and embodying the crisis ordinary.

Law is thus a punitive mechanism that can be used to enact indirect violence upon women. 'In today's economic climate, "peace" is something the profiteers of weak nations frequently enforce and "law" something they often employ to formalise their gains', Alexandra Kent (2016, p. 7) argues. Indeed, *Home SOS* has provided a nuanced analysis of the omissions and commissions that are a constituent part of intimate war being waged against women's bodies and the home at multiple and intermeshing scales. The book has thereby contributed to legal geography's renewed interest 'in the realities and materialities of law, including the ways in which they are experienced, and also the ways they are avoided, ignored, undermined, absent, oppressive or irrelevant' (Robinson and Graham 2018, p. 4).

Feminist legal geographies is a future avenue through which to find a better understanding of the 'law-space-power nexus' (Braverman et al. 2014, p. 18) that *Home SOS* has explored (Brickell and Cuomo 2019). As far back as 1992, Nicholas Blomley and Joel Bakan (1992) noted what they described as a 'surprising' neglect of legal ideology and legal discourse amongst critical geography work that had 'devoted considerable energy to examining the spatiality of structures that constrain human action and social consciousness, with special attention given to the politics of race, class and gender' (p. 665). While the legal geographies project went on to show 'often in granular detail, how unjust geographies are made and potentially un-made' (Delaney 2015, p. 268), the gendered spatialities and temporalities of law and (in)justice remain marked more by their absence than presence. A decade ago, Lisa Pruitt (2008, p. 338), also asserted that 'a more grounded and nuanced understanding of women's lived realities requires legal scholars to engage geography', yet the potential for dialogue between feminist geography, the legal geographies project, and (feminist) legal studies is still in its infancy. The book therefore has contributed to catalysing this dialogue moving forward.

The marking out of the indissolvable relations between law, space, and the workings of power across intimate and global scales is also something that *Home SOS* has developed through its focus on the extra-domestic home and the co-joining of domestic violence and forced eviction case studies. Its attention to the pluri-legal order of contemporary Cambodia has demonstrated the need for sensitivity to the multi-scalar coexistence of law of various kinds, with their own foundations of legitimacy, validity, power, and authority (von Benda-Beckmann and von Benda-Beckmann, 2014). Just as feminist legal studies seek to explore and place more centrally the multiple 'legal worlds' (Manji 1999) that women inhabit, feminist legal geographies must take an interlegal view of legal regimes and their dynamic interaction (Kotiswaran 2015; Valverde 2015) to unpack the 'power-geometries of time-space' (Massey 1999) within which women's (displaced) lives are lived. The book has not only shown how the intimate wars of domestic violence and forced eviction are mediated through law but how laws condition, and are conditioned by, the political order in which the crisis ordinary is created, sustained, and/or reckoned with. As a result, it has contributed to building a distinctive new development in human geography that demonstrates not only the spatial and temporal contingencies of law but also their significance and conflicted capacities to bring about positive change.

Conclusion

Positioning the home as a kaleidoscope through which to explore the interrelationships between gender, violence, and survival in crisis ordinary Cambodia, *Home SOS* has revealed the domestic sphere to be an ambiguous and non-linear space of brutalisation and coercion as well as creativity and agency in women's lives. It is at once a space of intimacy, comfort, and longevity, as well as chronic violence, estrangement, and truncation. As such, the home calls for an approach that reflects how 'dying and the ordinary reproduction of life are coexistent' (Berlant 2007, p. 102). The crisis ordinary has been mobilised in an original and grounded way as a critical pedagogy and imaginary through which to better comprehend and communicate women's experiences of survival tied to, and/or cast adrift from, home.

Usually thought of as separate entities, *Home SOS* has dared domestic violence and forced eviction to be understood as overwhelming ordinaries that are built into the infrastructures of the extra-domestic home. The book's attention to multiple crisis points – located in, and through, the spatialities and temporalities of home – challenges still commonly held views of the site as of niche rather than wide-ranging scholarly significance. I have evidenced how the home has both horizontal and vertical dimensions of spatial resonance: the horizontal encompassing the extra-domestic spaces of the street, parliament, the courts, and prison and the vertical extending to the global human rights regime and mutating geopolitical relationships. Across the domestic violence and forced eviction case studies, the flattening of these extra-domestic geographies have been at the core of government efforts to contain their 'fires' to the 'private' domain. The intellectual contribution of the book is therefore its reaffirmation and reprioritisation of the home as a political entity that is foundational to the concerns of human geography. Its stories provide a basis for thinking beyond extra-ordinary crisis events and working towards a discipline in which the ordinaries of violence and their differential impacts are foregrounded. These are vital if we are to make legible avoidable injustices and forged solidarities where capitalism and patriarchy would rather remain beyond closed doors.

Endnote

1 In a country of 15 million, it is estimated that 2 million Cambodians currently have a loan with a microfinance institution (Sahmakum Teang Tnaut and LICADHO 2019).

References

Ablin, D., and Hood, M. (1988). *The Cambodian Agony*. New York: M.E. Sharpe.

Abu-Lughod, L. (1990). The romance of resistance: Tracing transformations of power through Bedouin women. *American Ethnologist* 17 (1): 41–55.

Action Aid. (2010). Destined to fail? How violence against women is undoing development. https://www.actionaid.org.uk/sites/default/files/doc_lib/destined_to_fail.pdf (accessed 31 January 2019).

Adelman, M. (2004). The battering state: Towards a political economy of domestic violence. *Journal of Poverty* 8 (3): 45–64.

Adey, P. (2016). Emergency mobilities. *Mobilities* 11 (1): 32–48.

Adey, P. (forthcoming). Shoe: Towards a promiscuous politics of emergency evacuation mobility. *Environment and Planning D*.

Adey, P., Anderson, B., and Graham, S. (2015). Introduction: Governing emergencies: Beyond exceptionality. *Theory, Culture and Society* 32 (2): 3–17.

Agamben, G. (1998). *Homo Sacer: Sovereign Power and Bare Life*. Stanford, CA: Stanford University Press.

Ahmad, N., and Lahiri-Dutt, K. (2006). Engendering mining communities: Examining the missing gender concerns in coal mining displacement and rehabilitation in India. *Gender, Technology and Development* 10 (3): 313–339.

Ahmann, C. (2018). 'It's exhausting to create an event out of nothing': Slow violence and the manipulation of time. *Cultural Anthropology* 33 (1): 142–171.

Ahmetbeyzade, C. (2008). Gendering necropolitics: The juridical-political sociality of honor killings in Turkey. *Journal of Human Rights* 7 (3): 187–206.

Al-Ali, N., and Koser, K. (eds.) (2002). *New Approaches to Migration? Transnational Communities and the Transformation of Home*. London: Routledge.

Allad, T., and Thul, P.C. (2018). Election monitoring groups in Cambodia headed by PM's son, 'ambassador'. *Reuters* (20 July 2018). https://www.reuters.com/article/us-cambodia-election-monitors/election-monitoring-groups-in-cambodia-headed-by-pms-son-ambassador-idUSKBN1KA0FS (accessed 23 November 2018).

Home SOS: Gender, Violence, and Survival in Crisis Ordinary Cambodia, First Edition. Katherine Brickell.
© 2020 Royal Geographical Society (with the Institute of British Geographers). Published 2020 by John Wiley & Sons Ltd.

Allison, A. (2013). *Precarious Japan*. Durham, NC: Duke University Press.

Alves, J.A. (2014). From necropolis to blackpolis: Necropolitical governance and black spatial praxis in São Paulo, Brazil. *Antipode* 46 (2): 323–339.

Amnesty International. (2010). *Breaking the Silence: Sexual Violence in Cambodia*. London: Amnesty International.

Amnesty International. (2011). *Eviction and Resistance in Cambodia: Five Women Tell their Stories*. http://www.amnesty.org/en/library/info/ASA23/006/2011 (accessed 3 December 2018).

Amnesty International. (2012). *Housing is a Human Right*. https://www.amnesty.org.uk/files/youth_campaign_briefing_octnov12-forced_evictions.pdf (accessed 30 January 2019).

Amnesty International. (2015). *Taking to the Streets: Freedom of Peaceful Assembly in Cambodia*. London: Amnesty International.

Amnesty International. (2017). *Courts of Injustice: Suppressing Activism Through the Criminal Justice System in Cambodia*. London: Amnesty International.

Amnesty International. (2018). 'We are willing to risk everything to defend our homes and land'. https://www.amnesty.org/en/latest/campaigns/2018/10/we-are-willing-to-risk-everything-to-defend-our-homes-and-land/ (accessed 5 February 2019).

Anderson, B. (2015). What kind of thing is resilience? *Politics* 35 (1): 60–66.

Anderson, B. (2016). Governing emergencies: The politics of delay and the logic of response. *Transactions of the Institute of British Geographers* 41 (1): 14–26.

Anderson, B., Grove, K., Rickards, L., and Kearnes, M. (2019). Slow emergencies: Temporality and the racialized biopolitics of emergency governance. *Progress in Human Geography*. Online before print.

Anderson, K. (2005). Turning reconciliation on its head: Responding to sexual violence under the Khmer Rouge. *Seattle Journal for Social Justice* 3 (2): 785–832.

Anglin, M.K. (1998). Feminist perspectives on structural violence. *Identities* 5 (2): 145–151.

Ansell, N., Hadju, F., van Blerk, L., and Robson, E. (2018). 'My happiest time' or 'my saddest time'? The spatial and generational construction of marriage among youth in rural Malawi and Lesotho. *Transactions of the Institute of British Geographers* 43 (2): 184–199.

Anthony, K. (1997). Bitter homes and gardens: The meanings of home to families of divorce. *Journal of Architectural and Planning Research* 14: 1–19.

Arensen, L.J. (2012). Displacement, diminishment, and ongoing presence: The state of local cosmologies in northwest Cambodia in the aftermath of war. *Asian Ethnography* 71 (2): 159–178.

Arnold, D., and Chang, D.-O. (2017). Labor rights and trade unions in Cambodia. In: *The Handbook of Contemporary Cambodia* (eds. K. Brickell and S. Springer), 191–201. London: Routledge.

Ashe, V.H. (1988). *From Phnom Penh to Paradise*. London: Hodder & Stoughton Religious.

Asian Correspondent. (2016). Cambodian PM Hun Sen rides motorbike without helmet, gets fined. *Asian Correspondent* (24 June 2016). https://asiancorrespondent.com/2016/06/cambodian-pm-rides-motorbike-without-helmet-gets-fined/ (accessed 30 January 2019).

Asian Development Bank (ADB). (2012). *Cambodia Gender Analysis 2014*. Manila: ADB. https://www.adb.org/sites/default/files/institutional-document/33752/files/cambodia-country-gender-analysis.pdf (accessed 10 February 2019).

Asian Development Bank (ADB). (2014). *Cambodia Country Poverty Analysis*. Manila: ADB. https://www.adb.org/sites/default/files/institutional-document/151706/cambodia-country-poverty-analysis-2014.pdf (accessed 10 February 2019).

Asian Development Bank (ADB). (2015). *Promoting Women's Economic Empowerment in Cambodia*. Mandaluyong City, Philippines: ADB. https://www.adb.org/sites/default/files/publication/156499/promoting-womens-economic-empowerment.pdf (accessed 10 February 2019).

Askins, K. (2011). Geopolitics and activists: People, power and place. In: *The Ashgate Comparison to Critical Geopolitics* (eds. K. Dodds, M. Kuus, and J. Sharp), 527–542. Aldershot: Ashgate.

Aslam, A. (2017). *Ordinary Democracy: Sovereignty and Citizenship Beyond the Neoliberal Impasse*. Oxford: Oxford University Press.

Association of World Council of Churches Related Development Organisations in Europe (APRODEV). (2011). *Stolen Land, Stolen Future: A Report on Land Grabbing in Cambodia*. Brussels: APRODEV.

Astbury, J., and Walji, F. (2014). The prevalence of psychological costs of household violence by family members against women with disabilities in Cambodia. *Journal of Interpersonal Violence* 29 (17): 3127–3149.

Ayres, D.M. (2000). Tradition, modernity, and the development of education in Cambodia. *Comparative Education Review* 4 (4): 440–463.

Badgett, L., and Folbre, N. (1999). Assigning care: Gender norms and economic outcomes. *International Labour Review* 138 (3): 311–326.

Baeten, G., and Listerborn, C. (2015). Renewing urban renewal in Landskrona, Sweden: pursuing displacement through housing policies. *Geografiska Annaler: Series B, Human Geography* 97 (3): 249–261.

Bailey, M.M., and Shabazz, R. (2014). Gender and sexual geographies of blackness: New black cartographies of resistance and survival (Part 2). *Gender, Place and Culture* 21 (4): 449–452.

Baliga, A., and Chheng, N. (2017). A year of 'hell' in Prey Sar for Tep Vanny. *The Phnom Penh Post* (5 August 2017). https://www.phnompenhpost.com/national-post-depth/year-hell-prey-sar-tep-vanny (accessed 31 October 2018).

Bammer, A. (1992). Editorial. *New Formations* 17: vii–xi.

Banerjee, S.B. (2008). Necrocapitalism. *Organization Studies* 29 (12): 1541–1563.

Barash, D.P., and Webel, C. (2009). *Peace and Conflict Studies*. Thousand Oaks: Sage Publications.

Basso, A.R. (2016). Towards a theory of displacement atrocities: The Cherokee trail of tears, the herero genocide, and the Pontic Greek genocide. *Genocide Studies and Prevention: An International Journal* 10 (1): 5–29.

Basu, S. (2010). Judges of normality: Mediating marriage in the family courts of Kolkata, India. *Signs: Journal of Women in Culture* 37 (2): 469–492.

Bateman, M. (2017). *Post-War Reconstruction and Development in Cambodia and the Destructive Role of Microcredit*. 8th International Scientific Conference on 'Future World by 2050' at Pula, Croatia.

Bateman, M., Natarajan, N., Brickell, K., and Parsons, L. (2019). Descending into debt in Cambodia. *Made in China Journal* (January to March 2019). https://madeinchinajournal.com/2019/04/18/descending-into-debt-in-cambodia%EF%BB%BF%EF%BB%BF/ (accessed 1 September 2019).

Baviskar, A. (2009). Breaking homes, making cities: Class and gender in the politics of urban displacement. In: *Displaced by Development: Confronting Marginalization and Gender Injustice* (ed. L. Mehta), 59–81. New Delhi: Sage.

Baxter, R., and Brickell, K. (2014). For home unmaking. *Home Cultures* 11 (2): 133–143.

BBC (2018) One country's plan to solve the world's hidden health crisis. (BBC 25 October). http://www.bbc.com/future/story/20181023-nepals-plans-to-save-womens-lives (accessed 1 July 2019).

Beaugrand, C. (2011). Statelessness and administrative violence: Bidūns' survival strategies in Kuwait. *The Muslim World* 101: 228–250.

Beban, A., and Pou, S. (2014). *Human Security and Land Rights in Cambodia*. Phnom Penh: Cambodian Institute for Cooperation and Peace.

Beban, A., Sokbunthoeun, S., and Un, K. (2017) From force to legitimation: Rethinking land grabs in Cambodia. *Development and Change* 48 (3): 590–612.

Beban, A., and Work, C. (2014). The spirits are crying: Dispossessing land and possessing bodies in rural Cambodia. *Antipode* 46 (3): 593–610.

Becker, E. (1998). *When the War Was Over*. New York: Public Affairs.

Bell, T. (2008). Married Cambodian couple saw home in half after separation. *The Daily Telegraph* (9 October 2008). https://www.telegraph.co.uk/news/newstopics/howaboutthat/3159189/Married-Cambodian-couple-saw-home-in-half-after-separation.html (accessed 30 January 2019).

Benjamin, W. (1921). Critique of violence. In: *Reflections: Essays, Aphorisms, Autobiographical Writings* (ed. P. Demetz), 277–300. New York: Schocken Books.

Berger, J. (2007). *Hold Everything Dear: Dispatches on Survival and Resistance*. New York: Pantheon.

Berlant, L. (1998). Intimacy: A special issue. *Critical Inquiry* 24 (2): 281–288.

Berlant, L. (2007). Slow death (sovereignty, obesity, lateral agency). *Critical Inquiry* 33 (4): 754–780.

Berlant, L. (2011). *Cruel Optimism*. Durham, NC: Duke University Press.

Berlant, L. (2014). She's having an episode: Patricia Williams and the writing of damaged life. *Columbia Journal of Gender and Law* 27: 19–35.

Berlant, L. (2016). The commons: Infrastructures for troubling times. *Environment and Planning D* 34 (3): 393–419.

Berlant, L. (2018). Without exception: On the ordinariness of violence. Interview by Brad Evans. https://lareviewofbooks.org/article/without-exception-on-the-ordinariness-of-violence/ (accessed 1 February 2019).

Betlemidze, M. (2015). Mediatized controversies of feminist protest: FEMEN and bodies as affective events. *Women's Studies in Communication* 38 (4): 374–379.

Bhabha, H. (1994). *The Location of Culture*. New York: Routledge.

Bhuyan, R., Mell, M., Senturia, K., Sullivan, M., and Shiu-Thornton, S. (2005). 'Women must endure according to their karma': Cambodian immigrant women talk about domestic violence. *Journal of Interpersonal Violence* 20 (8): 902–921.

Biddulph, R. (2011). *Field Research Report Phase 1 of Gender Law and Economics at the Village Gate*. Gothenburg: Gothenburg Centre for Globalisation and Development.

Blomley, N. (2003) 'From "what?" to "so what?"': Law and geography in retrospect. In: *Law and Geography* (eds. J. Holder and C,. Harrison), 17–34, Oxford: Oxford University Press.

Blomley, N., and Bakan, J.C. (1992). Spacing out: Towards a critical geography of law. *Osgoode Hall Law Journal* 30 (3): 661–690.

Blunt, A., and Dowling, R.M. (2006). *Home*. London: Routledge.

Blunt, A., and Varley, A. (2004). Introduction: Geographies of home. *Cultural Geographies* 11: 3–6.

Bonneau, R. (2004). Professional views on the domestic violence draft law. GTZ Promotion of Women's Rights P3/D3.1.

Bosco, F. (2006). The madres de Plaza de Mayo and three decades of human rights activism: Embeddedness, emotions and social movements. *Annals of the Association of American Geographers* 96 (4): 342–365.

Bottomley, R. (2014). The role of civil society in influencing policy and practice in Cambodia. Report for Oxfam Novib. https://www.worldcitizenspanel.com/assets/The-Role-of-Civil-Society-in-Influencing-Policy-and-Practice-in-Cambodia.pdf (accessed 5 November 2017).

Bourdieu, P. (1987). The force of law: Toward a sociology of the juridical field. *Hastings Law Journal* 38: 805–853.

Bourdieu, P. (1991). *Language and Symbolic Power*. Cambridge, MA: Harvard University Press.

Bouvard, M. (1994). *Revolutionizing Motherhood: The Mothers of the Plaza de Mayo*. Wilmington, DE: Scholarly Resources.

Bowstead, J. (2015). Forced migration in the United Kingdom: Women's journeys to escape domestic violence. *Transactions of the Institute of British Geographers* 40 (3): 307–320.

Bowstead, J. (2017). Women on the move: Theorising the geographies of domestic violence journeys in England. *Gender, Place and Culture* 24 (1): 108–121.

Braidotti, R. (2011). *Nomadic Theory: The Portable Rosi Braidotti*. New York: Columbia University Press.

Braidotti, R. (2013). *The Posthuman*. Cambridge: Polity Press.

Braverman, I., Blomley, N., Delaney, D., and Kedar, A. (2014). Expanding the spaces of law. In: *The Expanding Spaces of Law: A Timely Legal Geography* (eds. N. Blomley, D. Delaney, and A. Kedar), 1–29. Stanford, CA: Stanford University Press.

Bregazzi, H., and Jackson, M. (2018). Agonism, critical political geography, and the new geographies of peace. *Progress in Human Geography* 42 (1): 72–91.

Brickell, K. (2007). *Gender Relations in the Khmer 'Home': Post-Conflict Perspectives*. PhD thesis. London School of Economics and Political Science.

Brickell, K. (2008). 'Fire in the House': Gendered experiences of drunkenness and violence in Siem Reap, Cambodia. *Geoforum* 39 (5): 1667–1675.

Brickell, K. (2011). 'We don't forget the old rice pot when we get the new one': Gendered discourses on ideals and practices of women in contemporary Cambodia. *Signs: Journal of Women in Culture and Society* 36 (2): 438–462.

Brickell, K. (2012a). Geopolitics of home. *Geography Compass* 6 (10): 575–588.

Brickell, K. (2012b). Mapping and doing critical geographies of home. *Progress in Human Geography* 36 (2): 225–244.

Brickell, K. (2012c). The 'stubborn stain' on development: Gendered meanings of housework (non)-participation in Cambodia. *The Journal of Development Studies* 47 (9): 1353–1370.

Brickell, K. (2013a). Cambodia's women activists are redefining the housewife. *The Guardian*, Comment is Free (2 April 2013). http://www.guardian.co.uk/commentisfree/2013/apr/02/cambodia-activists-housewife.

Brickell, K. (2013b). The real housewives of Cambodia. *The Telegraph* (4 April 2013). http://www.telegraph.co.uk/women/womens-life/9971064/The-real-housewives-of-Cambodia.html.

Brickell, K. (2014a). 'The whole world is watching': Intimate geopolitics of forced eviction and women's activism in Cambodia. *Annals of the Association of American Geographers* 104 (6): 1256–1272.

Brickell, K. (2014b). 'Plates in a basket will rattle': Marital dissolution and home 'unmaking' in contemporary Cambodia. *Geoforum* 51 (1): 262–272.

Brickell, K. (2015). Towards intimate geographies of peace? Local reconciliation of domestic violence in Cambodia. *Transactions of the Institute of British Geographers* 40 (3): 321–333.

Brickell, K. (2016). Gendered violences and rule of/by law in Cambodia. *Dialogues in Human Geography* 6 (2): 182–185.

Brickell, K. (2017a). Violence against women and girls in Cambodia. In: *The Handbook of Contemporary Cambodia* (eds. K. Brickell and S. Springer), 294–305. London: Routledge.

Brickell, K. (2017b). Clouding the judgment of domestic violence law: Victim blaming by institutional stakeholders in Cambodia. *Journal of Interpersonal Violence* 32 (9): 1358–1378.

Brickell, K., and Chant, S. (2010). 'The unbearable heaviness of being': Expressions of female altruism in Cambodia, Philippines, The Gambia and Costa Rica. *Progress in Development Studies* 10 (2): 145–159.

Brickell, K., and Cuomo, D. (2019). Feminist geolegality. *Progress in Human Geography* 43 (1): 104–122.

Brickell, K., and Maddrell, A. (2016). Gendered violences: The elephant in the room and moving beyond the elephantine. *Dialogues in Human Geography* 6 (2): 206–208.

Brickell, K., and Platt, M.W. (2015). Everyday politics of (in)formal marital dissolution in Cambodia and Indonesia. *Ethnos: Journal of Anthropology* 80 (3): 293–319.

Brickell, K., Prak, B., and Poch, B. (2014). *Domestic Violence Law: The Gap Between Legislation and Practice in Cambodia and What Can Be Done About It*. http://www.katherinebrickell.com/katherinebrickell/wp-content/uploads/2014/01/DV-Law-Prelim-Report-2014.pdf (accessed 1 February 2019).

Brickell, K., and Speer, J. (2020). Gendered and feminist approaches to displacement. In: *Handbook of Displacement* (eds. P. Adey, J. Bowstead, K. Brickell, V. Desai, M. Dolton, A. Pinkerton, and A. Siddiqi). London: Palgrave Macmillan.

Brickell, K., and Springer, S. (2017). Introduction to contemporary Cambodia. In: *The Handbook of Contemporary Cambodia* (eds. K. Brickell and S. Springer), 1–13. London: Routledge.

Brickell, K., Parsons, L., Natarajan, N., and Chann, S. (2018). *Blood Bricks: Untold Stories of Modern Slavery and Climate Change from Cambodia*. London: Royal Holloway, University of London. www.projectbloodbricks.org (accessed 1 February 2019).

Brinkley, J. (2011). *Cambodia's Curse: The Modern History of a Troubled Land*. New York: Public Affairs.

Briones-Vozmediano, E., Goicolea, I., Ortiz-Barreda, G.M., Gil-González, D., and Vives-Cases, C. (2014). Professionals' perceptions of support resources for battered immigrant women: Chronicle of an anticipated failure. *Journal of Interpersonal Violence* 29: 1006–1027.

Brown, I. (2000). *Cambodia: An Oxfam Country Profile*. Oxford: Oxfam GB.

Brown, G. (2015). Marriage and the spare bedroom: Exploring the sexual politics of austerity. *ACME: An International Journal for Critical Geographies* 14 (4): 975–988.

Brun, C. (2000). Making young displaced men visible. *Forced Migration Review* 10: 8–12.

Bruzzone, M. (2017). Respatializing the domestic: gender, extensive domesticity, and activist kitchenspace in Mexican migration politics. *Cultural Geographies* 24 (2): 247–263.

Brydolf-Horwitz, R. (2018). Embodied and entangled: Slow violence and harm via digital technologies. *Environment and Planning C*. Online before print.

Bugalski, N., and Pred, D. (2009). *Land Titling in Cambodia: Formalizing Inequality*. Phnom Penh: Bridges Across Borders Cambodia.

Buire, C., and Staeheli, L.A. (2017). Contesting the 'active' in active citizenship: Youth activism in Cape Town, South Africa. *Space and Polity* 21 (2): 173–190.

Burgos, S., and Ear, S. (2010). China's strategic interests in Cambodia: Influence and resources. *Asian Survey* 50 (3): 615–639.

Burgos, S., and Ear, S. (2013). *The Hungry Dragon: How China's Resource Quest is Re-Shaping the World*. Abingdon: Routledge.

Burns, C., and Daly, K. (2014). Responding to everyday rape in Cambodia: Rhetorics, realities and *somroh somruel*. *Restorative Justice: An International Journal* 2 (1): 64–84.

Butler, J. (1989). Foucault and the paradox of bodily inscriptions. *The Journal of Philosophy* 86 (11): 601–607.

Butler, J. (1990). *Gender Trouble: Feminist and the Subversion of Identity*. New York: Routledge.

Butler, J. (2004). *Precarious Life, the Powers of Mourning and Violence*. New York: Verso.

Butler, J. (2009). *Frames of War: When is Life Grievable?* London: Verso.

Butler, J. (2016). Rethinking vulnerability and resistance. In: *Vulnerability in Resistance* (eds. J. Butler, Z. Gambetti, and L. Sabsay), 12–27. Durham, NC: Duke University Press.

Butler, J., and Athanasiou, A. (2013). *Dispossession: The Performance in the Political*. Cambridge: Polity Press.

Butler, J., Gambetti, Z., and Sabsay, L. (2016). Introduction. In: *Vulnerability in Resistance* (eds. J. Butler, Z. Gambetti, and L. Sabsay), 1–11. Durham, NC: Duke University Press.

Bylander, M. (2015). Credit as coping: Rethinking microcredit in the Cambodian context. *Oxford Development Studies* 43 (4): 533–553.

Bylander, M. (2017). Micro-saturated: The promises and pitfalls of microcredit as a development solution. In: *The Handbook of Contemporary Cambodia* (eds. K. Brickell and S. Springer), 64–75. London: Routledge.

Cagna, P., and Rao, N. (2016). Feminist mobilisation for policy change on violence against women: Insights from Asia. *Gender and Development* 24 (2): 277–290.

Calkin, S. (2015). 'Tapping' women for post-crisis capitalism. *International Feminist Journal of Politics* 17 (4): 611–629.

Calvente, L.B.Y., and Smicker, J. (2017). Crisis subjectivities: resilient, recuperable, and abject representations in the new hard times. *Social Identities* 25 (2): 141–155.

Cambodian Center for Human Rights (CCHR). (2011). Fact sheet: Case study series: Boeung Kak. https://cchrcambodia.org/admin/media/factsheet/factsheet/english/CCHR%20Case%20Study%20Fact%20Sheet%20-%20Boeung%20Kak%20(ENG).pdf (accessed 30 January 2019).

Cambodian Center for Human Rights (CCHR). (2013). Wanted: Deputy Police Chief Phuong Malay. http://www.cchrcambodia.org/our_work/campaigns/impunity-2013/file/Case%20File%20-%20Phuong%20Malay.pdf (accessed 30 January 2019).

Cambodian Center for Human Rights (CCHR). (2016). Cambodia's women in land conflict. https://cchrcambodia.org/admin/media/report/report/english/2016_09_27_cchr_report_Cam_Women_in_Land_Conflict_ENG.pdf (accessed 30 January 2019).

Cappellini, B., Marilli, A., and Parsons, A. (2014). The hidden work of coping: Gender and the micro-politics of household consumption in times of austerity. *Journal of Marketing Management* 30 (15–16): 1597–1624.

Carr, H., Edgeworth, B., and Hunter, C. (2018). Introducing precaritisation: Contemporary understandings of law and the insecure home. In: *Law and the Precarious Home: Socio Legal Perspectives on the Home in Insecure Times* (eds. H. Carr, B. Edgeworth, and C. Hunter), 1–22. London: Bloomsbury.

Casolo, J., and Doshi, S. (2013). Domesticated dispossessions? Towards a transnational feminist geopolitics of development. *Geopolitics* 18 (4): 800–834.

Chakrya, K.S. (2012). Victim sickened by jibe. *The Cambodia Daily* (3 August 2012). https://www.phnompenhpost.com/national/victim-sickened-jibe (accessed 31 October 2018).

Chakrya, K.S., and Doyle, K. (2011). Dark new chapter in B Kak story. *The Phnom Penh Post*. http://www.phnompenhpost.com/national/dark-new-chapter-b-kak-story (accessed 2 November 2017).

Chakyra, K.S., and Worrell, S. (2013). A more personal B Kak dispute makes airwaves. *Phnom Penh Post* (8 July 2013). https://opendevelopmentcambodia.net/news/a-more-personal-b-kak-dispute-makes-airwaves/ (accessed 31 October 2018).

Chan, S. (2003) *Not Just Victims – Conversations with Cambodian Leaders in the United States*. Chicago, IL: University of Illinois Press.

Chandler, D. (1991). *The Tragedy of Cambodian History: Politics, War and Revolution since 1945*. New Haven, CT: Yale University Press.

Chandler, D. (1992). *Brother Number One: A Political Biography of Pol Pot*. Boulder, CO: Westview Press.

Chandler, D. (1996) *Facing the Cambodian Past- Selected Essays 1971–1994*. Chiang Mai, Thailand: Silkworm.

Chandler, D. (1999) *Voices from S-21: Terror and History in Pol Pot's Secret Prison*. Berkeley, CA: University of California Press.

Chant, S. (1997). *Women-Headed Households: Diversity and Dynamics in the Developing World*. London: Macmillan.

Chant, S. (2007). *Gender, Generation and Poverty: Exploring the 'Feminisation of Poverty' in Africa, Asia and Latin America*. Cheltenham: Edward Elgar.

Chap, C. (2018). David Chandler mulls new edition of his 'History of Cambodia' with more pessimistic ending. *VOA* 13 April. https://www.voacambodia.com/a/david-chandler-mulls-new-edition-of-his-history-of-cambodia-with-more-pessimistic-ending/4346864.html (accessed 30 January 2019).

Chatterton, P., and Pickerill, J. (2010). Everyday activism and transitions towards post-capitalist worlds. *Transactions of the Institute of British Geographers* 35 (4): 475–490.

Chattopadhyay, S. (2018). Violence on bodies: space, social reproduction and intersectionality. *Gender, Place and Culture* 25 (9): 1295–1304.

Chau-Pech Ollier, L., and Winter, T. (eds.) (2006). *Expressions of Cambodia: The Politics of Tradition, Identity and Change*. London: Routledge.

Chounaird, V. (1994). Geography, law and legal struggles: Which ways ahead? *Progress in Human Geography* 18 (4): 415–440.

Christensen, P. (2016). Spirits in Cambodian politics. http://voices.uni-koeln.de/2016-2/ spiritsincambodianpolitics (accessed 1 February 2019).

Ciorciari, J.D., and Heindel, A. (2014). *Hybrid Justice: The Extraordinary Chambers in the Courts of Cambodia*. Michigan: The University of Michigan Press.

Cockburn, C. (2004). The continuum of violence: A gender perspective on war and peace. In: *Sites of Violence: Gender and Conflict Zones* (eds. W. Giles and J. Hyndman), 24–44. Berkeley, CA: University of California Press.

Cockburn, C., and Zarkov, D. (eds.) (2002). *The Postwar Moment: Militaries, Masculinities and International Peacekeeping, Bosnia and The Netherlands*. London: Lawrence and Wishart.

Coffey, S.C. (ed.) (2018). *Seeking Justice in Cambodia: Human Rights Defenders Speak Out*. Melbourne: Melbourne University Press.

Cohen, E. (2012). Contesting discourses of blood in the 'red shirts' protests in Bangkok. *Journal of Southeast Asian Studies* 43 (2): 216–233.

COHRE. (2010). Briefing Paper: The impact of forced evictions on women. http://www. cohre.org/news/documents/briefing-paper-the-impact-of-forced-evictions-on-women (accessed 24 January 2014).

Coker, D. (2006). Restorative justice, Navajo peacemaking and domestic violence. *Theoretical Criminology* 10 (1): 67–85.

Coleman, M. (2009). What counts as the politics and practice of security, and where? Devolution and immigrant security after 9/11. *Annals of the Association of American Geographers* 99 (5): 904–913.

Collins, P.H. (1998). It's all in the family: Intersections of gender, race, and nation. *Hypatia: A Journal of Feminist Philosophy* 13 (3): 62–82.

Comaroff, J.L., and Comaroff, J. (2009). Reflections on the anthropology of law, governance, and sovereignty in a brave new world. In: *Rules of Law and Laws of Ruling: On the Governance of Law* (eds. F. von Benda-Beckmann, K. von Benda-Beckman, and J. Eckert), 31–60. Burlington: Ashgate.

Committee on Economic, Social and Cultural Rights. (1991). General comment No.4: The right to adequate housing (Sixth session) (art.11 (1) of the Covenant): The right to adequate housing. https://www.globalhealthrights.org/wp-content/uploads/2013/10/ CESCR-General-Comment-No.-4-The-Right-to-Adequate-Housing1.pdf (accessed 7 February 2019).

Committee on Economic, Social and Cultural Rights. (1997). General comment No. 7: The right to adequate housing (art. 11 (1) of the Covenant): Forced evictions. https:// tbinternet.ohchr.org/_layouts/treatybodyexternal/Download.aspx?symbolno=INT/ CESCR/GEC/6430&Lang=en (accessed 4 February 2019).

Committee on the Elimination of Discrimination Against Women (CEDAW). (1992). General Recommendation No. 19. http://www.un.org/womenwatch/daw/cedaw/recommendations/ recomm.htm (accessed 1 February 2019).

Committee on the Elimination of Discrimination Against Women (CEDAW). (2013). Concluding observations on the combined fourth and fifth periodic reports of Cambodia. 29 October CEDAW /C/KHM/CO/4-5*. https://tbinternet.ohchr.org/_ layouts/treatybodyexternal/Download.aspx?symbolno=CEDAW/C/KHM/CO/4-5 &Lang=En (accessed 1 February 2019).

Committee on the Elimination of Discrimination Against Women (CEDAW). (2018). Sixth periodic report submitted by Cambodia under article 18 of the Convention, due in 2017. 9 July 2018. CEDAW/C/KHM/6. https://tbinternet.ohchr.org/_layouts/treatybodyexternal/Download.aspx?symbolno=CEDAW%2fC%2fKHM%2f6&Lang=en (accessed 1 February 2019).

Confortini, C.C. (2006). Galtung, violence, and gender: The case for a peace studies/feminism alliance. *Peace and Change* 31 (3): 333–367.

Connell, J., and Grimsditch, M. (2017). Forced relocation in Cambodia. In: *The Handbook of Contemporary Cambodia* (eds. K. Brickell and S. Springer), 223–234. London: Routledge.

Cook, S. (ed.) (2006). *Genocide in Cambodia and Rwanda*. New Brunswick: Transactions Publishers.

Cook, S., and Pincus, J. (2014). Poverty, inequality and social protection in Southeast Asia: An introduction. *Journal of Southeast Asian Economies* 31 (1): 1–17.

Cornwall, A. (2002). *Making Spaces, Changing Places: Situating Participation in Development*. IDS Working Paper 170. Brighton, UK: Institute of Development Studies.

Coutin, S.B. (2016). *Exiled Home: Salvadoran Transnational Youth in the Aftermath of Violence*. Durham, NC: Duke University Press.

Coventry, L. (2017). Civil society in Cambodia: Challenges and contestations. In: *The Handbook of Contemporary Cambodia* (eds. K. Brickell and S. Springer), 53–63. London: Routledge.

Cover, R. (1986). Violence and the word. *Yale Law Journal* 95 (8): 1601–1630.

Cowen, D., and Smith, N. (2009). After geopolitics? From the geopolitical social to geoeconomics. *Antipode* 41 (1): 22–48.

Cowen, D., and Story, B. (2016). Intimacy and the everyday. In: *The Ashgate Research Companion to Critical Geopolitics* (eds. K. Dodds, M. Kuss, and J. Sharp), 341–358. London: Ashgate.

Cram, E.D. (2014). Review of Cruel Optimism by Lauren Berlant. *Rhetoric and Public Affairs* 17 (2): 371–374.

Cuomo, D. (2013). Security and fear: The geopolitics of intimate partner violence policing. *Geopolitics* 18: 856–874.

Cuomo, D., and Brickell, K. (2019). Feminist legal geographies. *Environment and Planning A*. Online before print.

Curley, M. (2018). Governing civil society in Cambodia: Implications of the NGO law for the 'rule of law'. *Asian Studies Review* 42 (2): 247–267.

Curthoys, A. (2008). Genocide in Tasmania. In: *Empire, Colony, Genocide: Conquest, Occupation and Subaltern Resistance in World History* (ed. D. Moses). New York and Oxford: Berghahn Books.

Curtis, G. (1998). *Cambodia Reborn?: The Transition to Democracy and Development*. Washington, DC: Brookings Institution Press.

Dabashi, H. (2012). La Vita Nuda: Baring bodies, bearing witness. *Al Jazeera* (January 23). https://www.aljazeera.com/indepth/opinion/2012/01/201212111238688792.html (accessed 1 September 2019).

D'Aoust, A.-M. (2018). A moral economy of suspicion: Love and marriage migration management practices in the United Kingdom. *Environment and Planning D* 36 (1): 40–59.

Daley, P. (1991). Gender, displacement and social reproduction. Settling Burundi refugees in Western Tanzania. *Journal of Refugee Studies* 4 (3): 248–266.

Dalla Costa, M. (1972). Women and the subversion of the community. In: *The Power of Women and the Subversion of the Community* (eds. M. Dalla Costa and S. James), 19–54. Bristol: Falling Wall Press.

Darling, J. (2017). Forced migration and the city: Irregularity, informality, and the politics of presence. *Progress in Human Geography* 41 (2): 178–198.

Das, V. (2007). *Life and Words: Violence and the Descent into the Ordinary*. Berkeley, CA: University of California Press.

Datta, A. (2016a). The intimate city: Violence, gender and ordinary life in Delhi slums. *Urban Geography* 37 (3): 323–342.

Datta, A. (2016b). *The Illegal City: Space, Law and Gender in a Delhi Squatter Settlement*. London: Routledge.

Davidson, M. (2009). Displacement, space and dwelling: Placing gentrification debate. *Ethics, Place and Environment: A Journal of Philosophy and Geography* 12 (2): 219–234.

Davies, T. (2019). Slow violence and toxic geographies: 'Out of sight' to whom? *Environment and Planning C: Politics and Space*, 1–19. Online before print.

Davies T., Isakjee, A., and Dhesi, S. (2017). Violent inaction: The necropolitical experience of refugees in Europe. *Antipode* 49 (5): 1263–1284.

Davis, E.W. (2016). *Deathpower: Buddhism's Ritual Imagination in Cambodia*. New York: Columbia University Press.

De Abreu, M.J.A. (2018). May Day supermarket: Crisis, impasse, medium. *Critical Inquiry* 44 (4): 745–765.

Deibler, S. (2017). Rape by any other name: Mapping the feminist legal discourse regarding rape conflict onto transitional justice in Cambodia. *American University International Law Review* 32 (2): 500–537.

Delaney, D. (2010). *The Spatial, the Legal and the Pragmatics of World-Making: Nomospheric Investigations*. London: Routledge.

Delaney, D. (2015). Legal geography II: Discerning injustice. *Progress in Human Geography* 40 (2): 267–274.

De Langis, T., Strasser, J., Kim, T., and Taing, S. (2014). *Like Ghost Changes Body: A Study on the Impact of Forced Marriage under the Khmer Rouge Regime*. Phnom Penh: Transcultural Psychosocial Organisation.

DeLuca, K.M. (1999). *Image Politics*. New York: Guilford Press.

De Mel, N. (2017). A grammar of emergence: Culture and the state in the post-tsunami resettlement of Burgher women of Batticaloa, Sri Lanka. *Critical Asian Studies* 49 (1): 73–91.

Derks, A. (2008). *Khmer Women on the Move – Migration and Urban Experiences in Cambodia*. Amsterdam: Dutch University Press.

Desmond, M. (2016). *Evicted: Poverty and Profit in the American City*. London: Penguin Books.

De Sousa Santos, B. (1987). Law: a map of misreading. Toward a postmodern conception of law. *Journal of Law and Society* 14 (3): 279–302.

Deth, S.U., Moldashev, K., and Bulut, S. (2017). The contemporary geopolitics of Cambodia: Alignments in regional and global contexts. In: *The Handbook of Contemporary Cambodia* (eds. K. Brickell and S. Springer), 17–28. London: Routledge.

De Verteuil, G., and Golubchikov, O. (2016). Can resilience be redeemed? *City* 20 (1): 143–151.

Di Certo, B. (2012). US hears Boeung Kak women. *The Phnom Penh Post* (16 July 2012). https://www.phnompenhpost.com/national/us-hears-boeung-kak-women (accessed 30 January 2019).

Diprose, R. (2009). Women's bodies between national hospitality and domestic biopolitics. *Paragraph* 32 (1): 69–86.

Dodds, K. (1996). The 1982 Falklands War and a critical geopolitical eye: Steve Bell and the if … cartoons. *Political Geography* 15: 571–592.

Dolan, C. (2002). Collapsing masculinities and weak states: A case study of Northern Uganda. In: *Masculinities Matter! Men, Gender and Development* (ed. F. Cleaver), 57–83. London: Zed Books.

Doshi, S., and Ranganathan, M. (2019). Towards a critical geography of corruption and power in late capitalism. *Progress in Human Geography* 43 (3): 436–457.

Dowling, E. (2016). Valorised but not valued? Affective remuneration, social reproduction and feminist politics beyond the crisis. *British Politics* 11 (4): 452–468.

Drori, G.S. (2005). United Nations' dedications: A world culture in the making? *International Sociology* 20 (2): 175–199.

Duffy, T. (1994). Toward a culture of human rights in Cambodia. *Human Rights Quarterly* 16 (1): 82–104.

Dunlap, C. (2001). Law and military interventions: preserving humanitarian values in 21st century conflict. Paper presented at the Humanitarian Challenges in Military Intervention Conference, Washington. http://people.duke.edu/~pfeaver/dunlap.pdf (accessed 10 April 2017).

Dwyer, M.B. (2014). Micro-geopolitics: Capitalizing security in Laos's golden quadrangle. *Geopolitics* 19: 377–405.

Ebihara, M. (1990). Revolution and reformation of Cambodian village culture. In: *The Cambodian Agony* (eds. B. Ablin and A. Hood). New York: M.E. Sharpe.

Ebihara, M. (1993). Beyond suffering: The recent history of a Cambodian village. In: *The Challenge of Reform in Indochina* (ed. B. Ljunggren), 149–166. Boston, MA: Harvard Institute for International Development.

Edwards, P. (2008). The Moral geology of the present: Structuring morality, menace and merit. In: *People of Virtue: Reconfiguring Religion, Power and Moral Order in Cambodia* (eds. A. Kent and D. Chandler), 213–240. Copenhagen: NIAS Press.

Eileraas, K. (2014). Sex(t)ing revolution, femen-izing the public square: Aliaa Magda Elmahdy, nude protest, and transnational feminist body politics. *Signs: Journal of Women in Culture and Society* 40 (1): 40–52.

El-Bushra, J. (2000). Transforming conflict: Some thoughts on a gendered understanding of conflict processes. In: *States of Conflict: Gender, Violence and Resistance* (eds. S. Jacobs, S. Jacobson, and J. Marchbank), 66–86. London: Zed Books.

Elias, J., and Roberts, A. (2018). Situating gender scholarship in IPE. In: *Handbook on the International Political Economy of Gender* (eds. J. Elias and A. Roberts), 1–22. Cheltenham: Edward Elgar.

Ellsberg, M., Arango, D.J., Morton, M., Gennari, F., Kiplesund, S., Contreras, M., and Watts, C. (2015). Prevention of violence against women and girls: What does the evidence say? *The Lancet* 385: 1555–1566.

Elson, D. (2002). Gender justice, human rights and neoliberal economic policies. In: *Gender Justice, Development and Rights* (eds. M. Molyneux and S. Shahra), 78–114. Oxford: Oxford University Press.

Eng, N., and Hughes, C. (2016). Local spaces of peace in Cambodia? In: *Post-Liberal Peace Transitions: Between Peace Formation and State Formation* (eds. O.P. Richmond and S. Pogodda), 160–178. Edinburgh: University of Edinburgh Press.

Eng, S., Szmodis, W., and Grace, K. (2017). Cambodian remarried women are at risk for domestic violence. *Journal of Interpersonal Violence*. Online before print.

Enloe, C. (1990). *Bananas, Beaches and Bases: Making Feminist Sense of International Politics*. Berkeley, CA: University of California Press.

Enloe, C. (2011). The mundane matters. *International Political Sociology* 5 (4): 447–450.

Ensor, M. (2017). Lost boys, invisible girls: Children, gendered violence and wartime displacement in South Sudan. In: *Gender, Violence, Refugees* (eds. S. Buckley-Zistel and U. Krause), 197–218. New York: Berghahn Books.

Erman, T., and Hatiboğlu, B. (2017). Rendering responsible, provoking desire: Women and home in squatter/slum renewal projects in the Turkish context. *Gender, Place and Culture* 24 (9): 1283–1302.

Ettlinger, N. (2007). Precarity unbound. *Alternatives: Global, Local, Political* 32 (3): 319–340.

Farha, L. (2018). UN Special Rapporteur on the Right to Housing. http://www.unhousingrapp.org (accessed 1 December 2018).

Faria, C. (2017). Towards a countertopography of intimate war: Contouring violence and resistance in a South Sudanese diaspora. *Gender, Place and Culture* 24 (4): 575–593.

Fauveaud, G. (2017). Real estate, productions, practices and strategies in contemporary Phnom Penh: An overview of social, economic, and political issues. In: *The Handbook of Contemporary Cambodia* (eds. K. Brickell and S. Springer), 212–222. London: Routledge.

Federici, S. (2004). *Caliban and the Witch: Women, The Body and Primitive Accumulation*. Brooklyn, NY: Autonomedia.

Federici, S. (2012). *Revolution at Point Zero: Housework, Reproduction, and Feminist Struggle*. Oakland, CA: PM Press.

Federici, S. (2014). Foreword. In: *Patriarchy and Accumulation on a World Scale: Women in the International Division of Labour* (ed. M. Mies), ix–xxiv. London: Zed Books.

Federici, S. (2018a). On reproduction as an interpretative framework for social/gender relations. *Gender, Place and Culture* 25 (9): 1391–1396.

Federici, S. (2018b). *Witches, Witch-Hunting and Women*. Oakland, CA: PM Press.

Feigenbaum, A. (2010). 'Now I'm a happy dyke!': Creating collective identity and queer community in Greenham women's songs. *Journal of Popular Music Studies* 22 (4): 367–388.

Fein, H. (1997). Genocide by attrition 1939–1993: The Warsaw ghetto, Cambodia, and Sudan: Links between human rights, health, and mass death. *Health and Human Rights* 2 (2): 10–45.

Fernandez, B. (2018). Dispossession and the depletion of social reproduction. *Antipode* 50 (1): 142–163.

Fernández Arrigoitia, M. (2017). Unsettling resettlements: Community, belonging and livelihood in Rio de Janeiro's Minha Casa Minha Vida. In: *Geographies of Forced Eviction: Dispossession, Violence, Resistance* (eds. K. Brickell, M. Fernández Arrigoitia, and A. Vasudevan), 71–96. Basingstole: Palgrave Macmillan.

Fineman, M.A. (2012). *At the Boundaries of Law*. London: Routledge.

Fitzpatrick, D. (2015). The legal design of land grabs: Possession and the state in post-conflict Cambodia. In: *Land Grabs in Asia: What Role for the Law?* (eds. C. Carter and A. Harding), 67–82. London: Routledge.

Flower, B.C.R. (2016). *Donor-Funded Titling and Urban Transition: A Case Study of the Land Management and Administration Programme (LMAP) in Phnom Penh, Cambodia*. PhD thesis. University College London.

Flower, B.C.R. (2019a). Legal geographies of neoliberalism: Market-oriented tenure reforms and the construction of an 'informal' urban class in post-socialist Phnom Penh. *Urban Studies* 56 (12): 2408–2425.

Flower, B.C.R. (2019b). Built on solid foundations? Assessing the links between city-scale land titling, tenure security and housing investment. *Planning Theory and Practice* 20 (3): 358–375.

Fluri, J.L. (2009). Geopolitics of gender and violence 'from below'. *Political Geography* 28: 259–265.

Forrest, R. (2015). The ongoing financialisation of home ownership – new times, new contexts. *International Journal of Housing Policy* 15 (1): 1–5.

Foucault, M. (1997). The birth of biopolitics. In: *Michel Foucault: Ethics, Subjectivity, and Truth* (ed. P. Rabinow), 73–80. New York: New Press.

Fowler, K.A., Gladden, R.M., Vagi, K.J., Barnes, J., and Frazier, L. (2015). Increase in suicides associated with home eviction and foreclosure during the United States housing crisis: Findings from 16 national violent death reporting system States, 2005–2010. *American Journal of Public Health* 105 (2): 311–316.

Fox O'Mahony, L., and Sweeney, J.A. (eds.) (2016). *The Idea of Home in Law: Displacement and Dispossession*. Ashgate: Ashgate.

Fraiman, S. (2017). *Extreme Domesticity: A View from the Margins*. New York: Columbia University Press.

Fraser, N. (2009). Feminism, capitalism, and the cunning of history. In: *Re-Framing the Transnational Turn in American Studies* (eds. W. Fluck, D.E. Pease, and J. Carlos), 374–390. Dartmouth: Dartmouth College Press.

Free the 15 (2015). Children of imprisoned activists call out to Michelle Obama. https://freethe15.wordpress.com/2015/03/17/children-of-imprisoned-activists-call-out-to-michelle-obama/ (accessed 31 October 2018).

Freedman, J. (2016). Sexual and gender-based violence against refugee women: A hidden aspect of the refugee 'crisis'. *Reproductive Health Matters* 24 (47): 18–26.

Frieson, K. (2011). *Cambodia Case Study: Evolution Toward Gender Equality*. Background paper for the WDR 2012.

Fryberg, S.A., and Eason, A.E. (2017). Making the invisible visible: Acts of commission and omission. *Current Directions in Psychological Science* 26 (6): 554–559.

Fullilove, M. (2004). *Root Shock: How Tearing Up City Neighbourhoods Hurts America, and What We Can Do About It*. New York: One World/Ballentine Books.

Fulu, E., and Miedema, S. (2015). Violence against women: Globalizing the integrated ecological model. *Violence Against Women* 21 (12): 1431–1455.

Galtung, J. (1969). Violence, peace and peace research. *Journal of Peace Research* 6 (3): 167–191.

Galtung, J. (1996). *Peace by Peaceful Means*. London: Sage.

Galtung, J. (2001). After violence, reconstruction, reconciliation, and resolution: Coping with visible and invisible effects of war and violence. In: *Reconciliation, Justice and Coexistence: Theory and Practice* (ed. M. Abu-Nimer), 3–24. Maryland: Lexington Books.

García-Lamarca, M., and Kaika, M. (2016). 'Mortgaged lives': The biopolitics of debt and housing financialisation. *Transactions of the Institute of British Geographers* 41: 313–327.

Gattrell, A. (2010). 'A frog in a well': The exclusion of disabled people from work in Cambodia. *Disability and Society* 25 (3): 289–301.

Geary, D., and Narim, K. (2013). Once united Boeng Kak protestors show discord. *The Cambodia Daily* (2 July 2013). https://saveboeungkak.wordpress.com/2013/07/02/once-united-boeng-kak-protesters-show-discord/ (accessed 31 October 2018).

Gender and Development Cambodia (GADC). (2003). *Paupers and Princelings: Youth Attitudes toward Gangs, Violence, Drugs and Theft*. Phnom Penh: GADC.

Gidley, R. (2019). *Illiberal Transitional Justice and the Extraordinary Chambers in the Courts of Cambodia*. London: Palgrave Macmillan.

Giles, W., and Hyndman, J. (2004). Introduction: Gender and conflict in a global context. In: *Sites of Violence: Gender and Conflict Zones* (eds. W. Giles and J. Hyndman), 3–23. Berkeley, CA: University of California Press.

Gill, R., and Orgad, S. (2018). The amazing bounce-backable woman: Resilience and the psychological turn in neoliberalism. *Sociological Research Online* 23 (2): 477–495.

Glassman, J. (2010). *Bounding the Mekong: The Asian Development Bank, China and Thailand*. Honolulu: University of Hawai'i.

Global Witness. (2018). Don't be fooled. There is nothing 'humanitarian' about Cambodia's dictatorship. https://www.globalwitness.org/en-gb/blog/dont-be-fooled-there-nothing-humanitarian-about-cambodias-dictatorship/ (accessed 2 November 2018).

Gordon, N. (2014). Human rights as a security threat: Lawfare and the campaign against human rights NGOs. *Law and Society Review* 48 (2): 311–343.

Gorman, C.S. (2019) Feminist legal archeology, domestic violence and the raced-gendered juridical boundaries of U.S. asylum law. *Environment and Planning A: Economy and Space* 51 (5): 1050–1067.

Gorman-Murray, A., McKinnon, S., and Dominey-Howes, D. (2014). Queer domicide? LGBT displacement and home loss in natural disaster impact, response and recovery. *Home Cultures* 11 (2): 237–262.

Gorvett, J. (2011). Cambodians evicted in 'land grab'. *The Guardian* (29 March). https://www.theguardian.com/world/2011/mar/29/cambodia-evictions-land-rights-gorvett (accessed 30 January 2019).

Gottesman, E. (2003) *Cambodia After the Khmer Rouge*. New Haven, CN: Yale University Press.

Graham, N. (2019). *Sheltering from Violence: Women's Experiences of Safe Houses in Cambodia*. PhD thesis. Royal Holloway, University of London.

Graham, N., and Brickell, K. (2019). Sheltering from domestic violence: Women's experiences of punitive safety and unfreedom in Cambodian safe shelters. *Gender, Place and Culture* 26 (1): 111–127.

Gray, H. (2018). The 'war'/'not-war' divide: Domestic violence in the Preventing Sexual Violence Initiative. *The British Journal of Politics and International Relations*. Online before print.

Green, O.H. (1980). Killing and letting die. *American Philosophical Quarterly* 17 (3): 195–204.

Green, N., and Bylander, M. (forthcoming). The coercive power of debt: Microfinance and land dispossession in Cambodia. *Sociology of Development*. Online before print.

Green, N., and Estes, J. (2019). Precarious debt: Microfinance subjects and intergenerational dependency in Cambodia. *Antipode*. Online before print.

Gregory, D. (2007). Vanishing points. In: *Violent Geographies* (eds. D. Gregory and A. Pred), 205–236. New York: Routledge.

Gunawardana, S.J. (2016). 'To finish, we must finish': Everyday practices of depletion in Sri Lankan export-processing zones. *Globalizations* 13 (6): 861–875.

Hafner-Burton, E.M., and Tsutsui, K. (2005). Human rights in a globalizing world: The paradox of empty promises. *American Journal of Sociology* 110 (5): 1373–1411.

Hagood Lee, S. (2006). *'Rice Plus' Widows and Economic Survival in Rural Cambodia*. New York: Routledge.

Hall, S.M. (2019). A very personal crisis: Family fragilities and everyday conjunctures within lived experiences of austerity. *Transactions of the Institute of British Geographers* 44: 479–492.

Hall, D., Hirsch, P., and Li, T.M. (2011). Introduction. In: *Powers of Exclusion: Land Dilemmas in South East Asia* (eds. D. Hall, P. Hirsch, and T.M. Li), 1–26. Honolulu: University of Hawaii Press.

Handley, E. (2018). Australian police investigating death threat against Kem Ley's widow. *Phnom Penh Post* (22 March 2018). https://www.phnompenhpost.com/national/australian-police-investigating-death-threat-against-kem-leys-widow (accessed 31 October 2018).

Haritaworn, J., Kuntsman, A., and Posocco, S. (2016). Introduction. In: *Queer Necropolitics* (eds. J. Haritaworn, A. Kuntsman, and S. Posocco), 1–28. London: Routledge.

Harker, C. (2012). Precariousness, precarity and family: Notes from Palestine. *Environment and Planning A* 44 (4): 849–865.

Harker, C. (2017). Debt space: Topologies, ecologies and Ramallah, Palestine. *Environment and Planning D: Society and Space* 35 (4): 600–619.

Harris, I.C. (2005). *Cambodian Buddhism: History and Practice*. Honolulu: University of Hawaii Press.

Harris, E., and Nowicki, M. (2018). Cultural geographies of precarity. *Cultural Geographies* 25 (3): 387–391.

Harris, E., Nowicki, M. and Brickell, K. (2018). On-edge in the impasse: Inhabiting the housing crisis as structure-of-feeling. *Geoforum* 101: 156–164.

Harvey, D. (2003). *The New Imperialism*. Oxford: Oxford University Press.

Hattendorf, J. and Tollerud, T.R. (1997). Domestic violence: Counselling strategies that minimize the impact of secondary victimization. *Perspectives in Psychiatric Care* 33 (1): 14–24.

Hayden, S. (2017). Michelle Obama, mom-in-chief: The racialized rhetorical contexts of maternity. *Women's Studies in Communication* 40 (1): 11–28.

Heise, L. (1998). Violence against women: An integrated ecological framework. *Violence Against Women* 4 (3): 262–290.

Hennings, A. (2019). The dark underbelly of land struggles: The instrumentalization of female activism and emotional resistance in Cambodia. *Critical Asian Studies* 51 (1): 103–119.

Heuveline, P., and Poch, B. (2006). Do marriages forget their past? Marital stability in post-Khmer Rouge Cambodia. *Demography* 43 (1): 99–125.

Heynen, N. (2006). 'But it's alright, Ma, it's life, and life only': Radicalism as survival. *Antipode* 38 (5): 916–929.

Him, C. (2000). *When Broken Glass Floats: Growing up under the Khmer Rouge*. New York: W.W. Norton & Company.

Hinton, A. (2002). Purity and contamination in the Cambodian genocide. In: *Cambodia Emerges from the Past: Eight Essays* (ed. J. Ledgerwood), 60–90. De Kalb: Northern Illinois University.

Hinton, A. (2005). *Why Did They Kill? Cambodia in the Shadow of Genocide*. Berkeley, CA: University of California Press.

Hinton, A. (2018). *The Justice Facade: Trials of Transition in Cambodia*. Oxford: Oxford University Press.

Hinton, A., Woolford, A., and Benvenuto, J. (2014). *Colonial Genocide in Indigenous North America*. Durham, NC: Duke University Press.

Hitchings, R. (2012). People can talk about their practices. *Area* 44 (1): 61–67.

Hochschild, R.A. (1983). *The Managed Heart: Commercialization of Human Feeling*. Berkeley, CA: University of California Press.

Hoffman, B., and Duschinski, H. (2014). Contestations over law, power and representation in Kashmir Valley. *Interventions* 16 (4): 501–530.

Holmes, S. (2003). Lineages of the rule of law. In: *Democracy and the Rule of Law* (eds. A. Przeworski and J.M. Maravall), 19–61. Cambridge: Cambridge University Press.

Holston, J. (2008). *Insurgent Citizenship: Disjunctions of Democracy and Modernity in Brazil*. Princeton, NJ: Princeton University Press.

Holston, J. (2009). Dangerous spaces of citizenship: Gang talk, rights talk, and rule of law in Brazil. *Planning Theory* 8 (1): 12–31.

hooks, bell. (1991). *Yearning: Race, Gender and Cultural Politics*. London: Turnaround.

Horn, D.M., and Parekh, S. (2018). Introduction to 'displacement'. *Signs: Journal of Women in Culture and Society* 43 (3): 503–514.

Horton, J., and Kraftl, P. (2009). Small acts, kinds words and 'not too much fuss': Implicit activisms. *Emotion, Space and Society* 2: 14–23.

Hoskyns, C., and Rai, S.M. (2007). Recasting the global political economy: Counting women's unpaid work. *New Political Economy* 12 (3): 297–317.

Hubbard, P., and Lees, L. (2018) The right to community? *City* 22(1): 8–25.

Hudson, B. (2002). Restorative justice and gendered violence: Diversion or effective justice. *British Journal of Criminology* 42: 616–634.

Hughes, C. (2003). *The Political Economy of the Cambodian Transition, 1991–2001*. London: Routledge.

Hughes, C. (2006) The politics of gifts: Tradition and regimentation in contemporary Cambodia. *Journal of Southeast Asian Studies* 37 (3): 469–489.

Hughes, R., and Elander, M. (2017). Justice in the past: The Khmer Rouge tribunal. In: *The Handbook of Contemporary Cambodia* (eds. K. Brickell and S. Springer), 42–52. London: Routledge.

Hughes, C., and Eng, C. (2019). Facebook, contestation and poor people's politics: Spanning the urban–rural divide in Cambodia? *Journal of Contemporary Asia* 49 (3): 365–388.

Hughes, C., and Un, K. (eds.) (2011). *Cambodia's Economic Transformation*. Copenhagen: NIAS Press.

Human Rights Watch (2014). Cambodia: New crackdown on protesters. 13 November 2014. https://www.hrw.org/news/2014/11/13/cambodia-new-crackdown-protesters (accessed 31 October 2018).

Human Rights Watch (2015). At your own risk: Reprisals against critics of World Bank Group Projects. https://www.hrw.org/sites/default/files/reports/worldBank0615_4Up.pdf (accessed 1 February 2019).

Human Rights Watch (2017a). *Cambodia – Events of* 2017. https://www.hrw.org/world-report/2018/country-chapters/cambodia (accessed 23 November 2018).

Human Rights Watch (2017b). Cambodia: Quash conviction of rights land activist. 7 August 2017. https://www.hrw.org/news/2017/08/07/cambodia-quash-conviction-rights-land-activist (accessed 1 February 2019).

Hunt, L. (2014). Spotlight on Cambodian government brutality. *The Diplomat* (5 May). https://thediplomat.com/2014/05/spotlight-on-cambodian-government-brutality/ (accessed 1 February 2019).

Huseman, J., and Short, D. (2012). 'A slow industrial genocide': Tar sands and the indigenous peoples of northern Alberta. *The International Journal of Human Rights* 16 (1): 216–237.

Hyndman, J. (2001). Towards a feminist geopolitics. *The Canadian Geographer* 42 (2): 210–222.

Hyndman, J. (2019). Unsettling feminist geopolitics: Forging feminist political geographies of violence and displacement. *Gender, Place and Culture* 26 (1): 3–29.

ICNL. (2016) Civic Freedom Monitor: Cambodia. http://www.icnl.org/research/monitor/cambodia.html (accessed 30 January 2019).

IFEX. (2013). Cambodian activist wants to hear from you after a year in jail. https://www.ifex.org/cambodia/2013/06/17/photos4bopha/ (accessed 31 October 2018).

IFEX. (2014). Temporary arrest of Yorm Bopha and 4 other land activists in Cambodia; demonstrations banned in Phnom Penh. http://www.ifex.org/cambodia/2014/01/06/bopha_arrested/ (accessed 31 October 2018).

Inclusive Development International. (2012). *Cambodia: ADB and Australia-Financed Railway Project*. https://www.inclusivedevelopment.net/campaign/cambodia-adb-and-australia-financed-railway-project/ (accessed 30 January 2019).

International Labor Organisation and National Institute of Statistics. (2013). *Cambodia - Labor Force and Child Labor Survey 2012*. https://www.ilo.org/asia/WCMS_230721/lang--en/index.htm (accessed 5 October 2019).

Inwood, J., and Tyner, J. (2011). Geography's pro-peace agenda: an unfinished project. *ACME: An International E-Journal for Critical Geographies* 10 (3): 442–457.

Isin, E.F. (2008). Theorising acts of citizenship. In: *Acts of Citizenship* (eds. E. Isin and G. Neilsen), 15–44. London: Zed Books.

Isin, E.F. (2009). Citizenship in flux: The figure of the activist citizen. *Subjectivity* 29: 367–388.

Isin, E.F., and Rygiel, K. (2007). Abject spaces: Frontiers, zones, camps. In: *The Logics of Biopower and the War on Terror* (eds. E. Dauphinee and C. Masters), 181–203. New York: Palgrave Macmillan.

Itagaki, L.M. (2013). Crisis temporalities: States of emergency and the gendered-sexualized logics of Asian American women abroad. *Feminist Formations* 25 (2): 195–219.

Izumi, K. (2007). Gender-based violence and property grabbing in Africa: a denial of women's liberty and security. *Gender and Development* 15 (1): 11–23.

Jackson, K. (1989). *Cambodia: 1975–1978: Rendezvous with Death*. Princeton, NJ: Princeton University Press.

Jackson, C. (2012). Introduction: Marriage, gender relations and social change. *Journal of Development Studies* 48 (1): 1–9.

Jackson, W., and Monkolransey, M. (2015). How serious are Cambodia's land rights protesters about their curses? *Phnom Penh Post* (8 August). https://www.phnompenhpost.com/post-weekend/how-serious-are-cambodias-land-rights-protesters-about-their-curses (accessed 1 February 2019).

Jackson, W., and Turton, S. (2016). Clinton notified of Boeung Kak release, emails show. *The Phnom Penh Post* (23 March). https://www.phnompenhpost.com/national/clinton-notified-boeung-kak-release-emails-show (accessed 31 October 2018).

Jacobsen, T.A. (2003). *Threads in a Sampot: A History of Women and Power in Cambodia.* PhD thesis. University of Queensland.

Jacobsen, T.A. (2008). *Lost Goddesses: The Denial of Female Power in Cambodian History.* Copenhagen: NIAS Press.

Jacobsen, T.A. (2012). Being *broh*: The good, the bad, and the successful man in Cambodia. In: *Men and Masculinities in Southeast Asia* (eds. M. Ford and L. Lyons). London: Routledge.

Jegannathan, B., Dahlblom, K., and Kullgren, G. (2014), 'Plue plun' male, 'kath klei' female: Gender differences in suicidal behavior as expressed by young people in Cambodia. *International Journal of Culture and Mental Health* 7 (3): 326–338.

Johnston, L. (2018). Gender and sexuality III: Precarious places. *Progress in Human Geography* 42 (6): 928–936.

Jones, C.A. (2016). Lawfare and the juridification of late modern war. *Progress in Human Geography* 40 (2): 221–239.

Jones, C.A., and Smith, M.D. (2015). War/law/space: Notes toward a legal geography of law. *Environment and Planning D* 33: 581–591.

Jupp, E. (2017). Home space, gender and activism: The visible and the invisible in austere times. *Critical Social Policy* 37 (3): 348–366.

Juran, L. (2012). The gendered nature of disasters: Women survivors in post-tsunami Tamil Nadu. *Indian Journal of Gender Studies* 19 (1), 1–29.

Kaag, M., and Zoomers, A. (2014). *The Global Land Grab: Beyond the Hype.* London: Zed Books.

Kabeer, N. (1997). Tactics and trade-offs: Revisiting the links between gender and poverty. *IDS Bulletin* 28 (3): 1–13.

Kabeer, N. (1999). Resources, agency, achievements: Reflections on the measurement of women's empowerment. *Development and Change* 30: 435–464.

Kanngieser, A. (2013). *Experimental Politics and the Making of Worlds.* London: Routledge.

Kaplan, C. (2001). Hillary Rodham Clinton's orient: Cosmopolitan travel and global feminist subjects. *Meridians* 2 (1): 219–240.

Katz, C. (2001). Vagabond capitalism and the necessity of social reproduction. *Antipode* 33 (4): 709–728.

Katz, C. (2004). *Growing Up Global: Economic Restructuring and Children's Everyday Lives.* Minneapolis, MN: University of Minnesota Press.

Kelly, C. (2018). *A Cambodian Spring* [film]. https://acambodianspring.com.

Kent, A. (2011a). Sheltered by *dhamma*: Reflecting on gender, security and religion in Cambodia. *Journal of Southeast Asian Studies* 42 (2): 193–209.

Kent, A. (2011b). Global change and moral uncertainty: Why do Cambodian women seek refuge in Buddhism? *Global Change, Peace and Security* 23 (3): 405–419.

Kent, A. (2016). Conflict continues: Transitioning into a battle for property in Cambodia today. *Journal of Southeast Asian Studies* 47 (1): 3–23.

Khmer Times. (2016). Hun Sen fined for not wearing helmet. 23 June 2016. https://www.khmertimeskh.com/news/26446/hun-sen-fined-for-not-wearing-helmet/ (accessed 31 October 2018).

Kiernan, B. (2002). Introduction: Conflict in Cambodia, 1945–2002. *Critical Asian Studies* 34 (4): 483–495.

Kiernan, B. (2003). Research note: The demography of genocide in Southeast Asia. *Critical Asian Studies* 35 (4): 585–597.

Kiernan, B. (2008). *The Pol Pot Regime*. New Haven, CN: Yale University Press.

Kiernan, B., and Boua, C. (1982). *Peasants and Politics in Kampuchea, 1942–1981*. London: Zed Books.

Kijewski, L. (2018). Cambodia rights activist freed from jail after pardon. Aljazeera 20 August. https://www.aljazeera.com/news/2018/08/cambodia-rights-activist-freed-jail-pardon-180820163915085.html (accessed 2 November 2018).

Kijewski, L., and Sineat, Y. (2018). In Cambodia's courts, it's a man's world – with the effects felt by female employees and victims alike. *Phnom Penh Post* (23 February). https://www.phnompenhpost.com/national-post-depth/cambodias-courts-its-mans-world-effects-felt-female-employees-and-victims-alike (accessed 1 February 2019).

Kirby, P. (2015). Ending sexual violence in conflict: The preventing sexual violence initiative and its critics. *International Affairs* 91 (3): 457–472.

Kittrie, G.F. (2016) *Lawfare: Law as a Weapon of War*. Oxford: Oxford University Press.

Kohn, L. (2010). What's so funny about peace, love and understanding? Restorative justice as a new paradigm for domestic violence intervention. *Seton Hall Law Review* 40: 517–594.

Kotiswaran, P. (2015). Valverde's chronotopes of law: Reflections on an agenda for socio-legal studies. *Feminist Legal Studies* 23: 353–359.

Kraftl, P., and Horton, J. (2008). Spaces of every-night life: For geographies of sleep, sleeping and sleepiness. *Progress in Human Geography* 32 (4), 509–524.

Kry, S. (2014). The Boeung Kak development project: For whom and for what? Poor land development practices as a challenge for building sustainable peace in Cambodia. *Cambodia Law and Policy Journal* 3: 11–44.

Kumar, K., Baldwin, H., and Benjamin, J. (2001). Profile: Cambodia. In: *Women and Civil War: Impact, Organizations and Action* (ed. K. Kumar), 39–47. Boulder, CO and London: Lynne Rienner.

Lafreniere, B. (2000). *Music Through the Dark: A Tale of Survival in Cambodia*. Honolulu: University of Hawai'i.

Laing, L. (2017). Secondary victimization: Domestic violence survivors navigating the family law system. *Violence Against Women* 23 (11): 1314–1335.

Lala, S., and Straussner, A. (2001). *Ethnocultural Factors in Substance Abuse Treatment*. New York: Guilford Press.

Laliberté, N. (2015). Geographies of human rights: Mapping responsibility. *Geography Compass* 9 (2): 57–67.

Lamb, V., Schoenberger, L., Middleton, C., and Un, B. (2017). Gendered eviction, protest and recovery: A feminist political ecology engagement with land grabbing in rural Cambodia. *Journal of Peasant Studies* 44 (6): 1215–1234.

Lamble, S. (2013). Queer necropolitics and the expanding carceral state: Interrogating sexual investments in punishment. *Law Critique* 24: 229–253.

Lancione, M. (2017). Revitalising the uncanny: Challenging inertia in the struggle against forced evictions. *Environment and Planning D: Society and Space* 35 (6): 1012–1032.

Lancione, M. (2019a). Radical housing: On the politics of dwelling as difference. *International Journal of Housing Policy*. Online before print.

Lancione, M. (2019b). The politics of embodied urban precarity: Roma people and the fight for housing in Bucharest, Romania. *Geoforum* 101: 182–191.

Latham, A. (2002). Warfare transformed: A Braudelian perspective on the 'revolution in military affairs'. *European Journal of International Relations* 8 (2): 231–266.

Lawreniuk, S. (2019). Intensifying political geographies of authoritarianism: Towards an anti-geopolitics of garment workers struggles in neoliberal Cambodia. *Annals of the Association of American Geographers*. Online before print.

Lawreniuk, S., and Parsons, L. (2018). For a few dollars more: Towards a translocal mobilities of labour activism in Cambodia. *Geoforum* 92: 26–35.

Lawrie, E.W., and Shaw, I.G.R. (2018). Violent conditions: The injustices of being. *Political Geography* 65: 8–16.

Ledgerwood, J. (1990). *Changing Khmer Conceptions of Gender: Women, Stories and the Social Order*. PhD thesis. Cornell University.

Ledgerwood, J. (1994). Gender symbolism and culture change: Viewing the virtuous woman in the Khmer story 'Mea Yeong'. In: *Culture Since 1975: Homeland and Exile* (eds. M. Ebihara, C. Mortland, and J. Ledgerwood), 199–128. Ithica, NY: Cornell University Press.

Ledgerwood, J. (1995). Khmer kinship: The matriliny/matriarchy myth. *Journal of Anthropological Research* 51 (3): 247–261.

Ledgerwood, J. (1996). Politics and gender: Negotiating concepts of the ideal woman in present day Cambodia. *Asia Pacific Viewpoint* 32 (136), 139–152.

Ledgerwood, J. (2003). *Cambodian Recent History and Contemporary Society: An Introductory Course*. http://www.seasite.niu.edu/khmer/ledgerwood) (accessed April 2003).

Ledgerwood, J. (2012). Buddhist ritual and the reordering of social relations in Cambodia. *South East Asia Research* 20 (2): 191–205.

Lees, L., Annunziata, S., and Rivas-Alonso, C. (2018). Resisting planetary gentrification: The value of survivability in the fight to stay put. *Annals of the American Association of Geographers* 108 (2): 346–355.

Lemkin, R. (1944). *Axis Rule in Occupied Europe: Analysis, Proposals for Redress*. Washington, DC: Carnegie Endowment for International Peace.

Levine, P. (2010). *Love and Dread in Cambodia: Weddings, Births, and Ritual Harm under the Khmer Rouge*. Honolulu: University of Hawai'i Press.

Levy, A., and Scott-Clark, C. (2008). Country for sale. https://www.theguardian.com/world/2008/apr/26/cambodia (accessed 14 November 2018).

Lewis, R., Dobash, R., and Cavanagh, K. (2001). Law's progressive potential: The value of engagement with the law for domestic violence. *Social and Legal Studies* 10 (1): 105–130.

Li, T.M. (2010). To make life or let die? Rural dispossession and the protection of surplus populations. *Antipode* 41: 66–93.

LICADHO. (2012). Another mother from Boeung Kak imprisoned. http://www.licadho-cambodia.org/reports/files/173Free15+AI-BophaProfile-English.pdf (accessed 31 October 2018).

LICADHO. (2014a). *2014 Brings a New Wave of Cambodian Land Conflicts*. https://www.licadho-cambodia.org/press/files/342LICADHOPRLandConflictd2014-English.pdf (accessed 1 February 2019).

LICADHO. (2014b). *'Good Wives': Women Land Campaigners and the Impact of Human Rights Activism*. Phnom Penh: LICADHO.

LICADHO. (2017). *No Punishment, No Protection: Cambodia's Response to Domestic Violence*. Phnom Penh: LICADHO.

LICADHO Canada. (2011). Pushed to the edge: The death of a Boeung Kak lake activist [video]. https://www.youtube.com/watch?v=a82OMhG48Gc (accessed 2 November 2017).

Lifton, R.J. (1991). *Death in Life: Survivors of Hiroshima*. New York: North Carolina Press.

Lila, M., Gracia, E., and Gracia, F. (2013). Ambivalent sexism, empathy and law enforcement attitudes towards partner violence against women among male police officers. *Psychology, Crime and Violence* 19: 907–919.

Lilja, M. (2008). *Power, Resistance and Women Politicians in Cambodia: Discourses of Emancipation*. Copenhagen: NIAS Press.

Lilja, M. (2014). *Resisting Gender Norms: Civil Society, the Judicial and Political Space in Cambodia*. Farnham: Ashgate.

Lilja, M. (2017). Layer-cake figurations and hide-and-show resistance in Cambodia. *Feminist Review* 117 (1): 131–147.

Lister, R. (1997). Citizenship: Towards a feminist synthesis. *Feminist Review* 57: 28–48.

Lister, R. (2003). *Citizenship: Feminist Perspectives*. Basingstoke: Palgrave Macmillan.

Little, J. (2019). Violence, the body and the spaces of intimate war. *Geopolitics*. Online before print.

Liu, H.-Y. (2015). *Law's Impunity: Responsibility and the Modern Private Military Company*. London: Bloomsbury.

Longhurst, R. (2016). Semi-structured interviews and focus groups. In: *Key Methods in Geography* (eds. N. Clifford, M. Cope, T. Gillespie, and S. French), 143–156. London: Sage.

Loyd, J.M. (2012). Geographies of peace and antiviolence: Peace and antiviolence. *Geography Compass* 6 (8): 477–489.

Luco, F. (2002). *Between a Tiger and a Crocodile: Management of Local Conflicts in Cambodia: An Anthropological Approach to Traditional and New Practices*. Phnom Penh: UNESCO.

Lyons, T., Krusi, A., Pierre, L., Small, W., and Shannon, K. (2017). The impact of construction and gentrification on an outdoor trans sex work environment: Violence, displacement and policing. *Sexualities* 20 (8): 881–903.

MacDonald, D., and Logan, T. (2016). Guest editors' introduction: Genocide studies, colonization, and indigenous peoples. *Genocide Studies and Prevention* 10 (1): 2–4.

Macgregor Wise, J. (2000). Home: Territory and identity. *Cultural Studies* 14 (2): 295–310.

MacKinnon, D., and Derickson, K.D. (2012). From resilience to resourcefulness: A critique of resilience policy and activism. *Progress in Human Geography* 37 (2): 253–270.

MacLeod, G., and McFarlane, C. (2014). Introduction: Grammars of urban injustice. *Antipode* 46 (4): 857–873.

Madden, D., and Marcuse, P. (2016). *In Defense of Housing: The Politics of Crisis*. London: Verso.

Maddrell, A., and Sidaway, J.D. (eds) (2016). *Deathscapes: Spaces for Death, Dying, Mourning and Remembrance*. London: Routledge.

Mader, P. (2015). *The Political Economy of Microfinance*. London: Palgrave Macmillan.

Madge, C., Raghuram, P., Skelton, T., Willis, K., and Williams, J. (1997). Methods and methodologies in feminist geographies: Politics, practices and power. In: *Feminist Geographies: Explorations in Diversity and Difference* (eds. Women and Geography Study Group), 86–111. London: Routledge.

Maharawal, M.M., and McElroy, E. (2018). The Anti-Eviction Mapping Project: Counter mapping and oral history toward Bay Area housing justice. *Annals of the American Association of Geographers* 108 (2): 380–389.

Majidi, N., and Hennion, C. (2014). Resilience in displacement? Building the potential of Afghan displaced women. *Journal of Internal Displacement* 4 (1): 77–91.

Manchanda, R. (2004). Gender conflict and displacement: Contesting 'infantilisation' of forced migrant women. *Economic and Political Weekly* 39 (37): 4179–4186.

Manji, A.S. (1999). Imagining women's 'legal world': Towards a feminist theory of legal pluralism in Africa. *Social and Legal Studies* 8 (4): 435–455.

Marston, J. (2002). Democratic Kampuchea and the idea of modernity. In: *Cambodia Emerges from the Past: Eight Essays* (ed. J. Ledgerwood), 38–59. De Kalb: Northern Illinois University.

Masco, J. (2017). The crisis in crisis. *Current Anthropology* 58 (15): S65–S76.

Massey, D. (1994). *Space, Place and Gender*. Minneapolis, MN: University of Minnesota Press.

Massey, D. (1999). Imagining globalization: Power-geometries of time–space. In: *Global Futures. Explorations in Sociology* (eds. A. Brah, M.J. Hickmanm, and M.M. Ghaill), 27–44. London: Palgrave Macmillan.

Mayblin, L., Wake, M., and Kazemi, M. (2019). Necropolitics and the slow violence of the everyday: Asylum seeker welfare in the postcolonial present. *Sociology* 1–17. Online before print.

Mbembé, A. (2003). Necropolitics. *Public Culture* 15 (1): 11–40.

Mbembé, A. (2019). *Necropolitics*. Durham, NC: Duke University Press.

Mbembé, A., and Roitman, J. (1995). Figures of the subject in times of crisis. *Public Culture* 7: 323–352.

McCargo, D. (2005). Cambodia: Getting away with authoritarianism? *Journal of Democracy* 16 (4): 98–112.

McCarthy, S., and Un, K. (2017). The evolution of rule of law in Cambodia. *Democratization* 24 (1): 100–118.

McDonnell, M.A., and Moses, A.D. (2005). Raphael Lemkin as historian of genocide in the Americas. *Journal of Genocide Research* 7 (4): 504–505.

McDowell, L. (2001). 'It's that Linda again': Longitudinal research with young men. *Ethics, Place and Environment* 4 (2): 87–100.

McDowell, L. (2007). *Gender, Identity and Place: Understanding Feminist Geographies*. Minneapolis, MN: University of Minnesota Press.

McDowell, L. (2016). Reflections on feminist economic geography: Talking to ourselves? *Environment and Planning A.* 48 (10): 2093–2099.

McFarlane, C. (2011). The city as assemblage: Dwelling and urban space. *Environment and Planning D: Society and Space* 29: 649–671.

McGarvey, A. (2005). National Assembly approves domestic violence law. *Phnom Penh Post* (23 September 2005). http://www.phnompenhpost.com/national/national-assembly-approves-domestic-violence-law (accessed 30 January 2019).

McGinn, C. (2013). *'Every day is difficult for my body and my heart.' Forced Evictions in Phnom Penh, Cambodia: Women's Narratives of Risk and Resilience*. PhD thesis. Columbia University.

McIntyre, K. (1996). Geography as destiny: Cities, villages and Khmer Rouge orientalism. *Comparative Studies in Society and History* 38 (4): 730–758.

McKinnon, S.L. (2016). Necropolitical voices and bodies in the rhetorical reception of Iranian women's asylum claims. *Communication and Critical/Cultural Studies* 13 (3): 215–231.

McLinden Nuijen, M., Prachvuthy, M., and Van Westen, G. (2004). 'Land grabbing' in Cambodia: Land rights in a post-conflict sitting. In: *The Global Land Grab: Beyond the Hype* (eds. M. Kaag and A. Zoomers), 152–169. London: Zed Books.

McNeilly, K. (2018). Are rights out of time? International human rights law, temporality, and radical social change. *Social and Legal Studies*, 1–22. Online before print.

Mech, D., and Woods, B. (2013). Tep Vanny – From Boeng Kak protester to globe-trotting advocate. *The Cambodia Daily* (8 October). https://www.cambodiadaily.com/archives/tep-vanny-from-boeng-kak-protests-to-globe-trotting-advocate-44583/ (accessed 31 October 2018).

Meertens, D., and Segura-Escobar, N. (1996). Uprooted lives: Gender, violence and displacement in Columbia. *Singapore Journal of Tropical Geography* 17 (2): 165–178.

Mehrvar, M., Sore, K.C., and Sambath, M. (2008). Women's perspectives: A case study of systematic land registration: Cambodia. Phnom Penh: Gender and Development for Cambodia and Heinrich Böll Stiftung Cambodia.

Mehta, L. (ed.) (2009). *Displaced by Development: Confronting Marginalization and Gender Injustice*. Los Angeles, CA: Sage.

Menjívar, C., and Walsh, S.D. (2017). The architecture of feminicide: The state, inequalities, and everyday gender violence in Honduras. *Latin American Research Review* 52 (2): 221–240.

Merry, S.E. (1995). Resistance and the cultural power of law. *Law and Society Review* 29 (1): 11–26.

Merry, S.E. (2003). Rights talk and the experience of law: Implementing women's human rights to protection from violence. *Human Rights Quarterly* 25 (2): 343–381.

Merry, S.E. (2006). *Human Rights and Gender Violence: Translating International Law into Local Justice*. Chicago, IL: University of Chicago Press.

Merry, S.E. (2009). *Gender Violence: A Cultural Perspective*. Malden, MA: Wiley-Blackwell.

Merry, S.E., and Levitt, P. (2017). The vernacularization of women's human rights. In: Hopgood, S., Snyder, J., Vinjamuri, L. (eds.) *Human Rights Futures* (eds. S. Hopgood, J. Snyder, and L. Vinjamuri), 213–236. Cambridge: Cambridge University Press.

Meta, K., and Kijewski, L. (2018). Future uncertain for women's radio station. *Phnom Penh Post* 18 January. Available from: https://www.phnompenhpost.com/national/future-uncertain-womens-radio-station (accessed 31 October 2018).

Meth, P. (2003). Rethinking the domus in domestic violence: homelessness, space and domestic violence in South Africa. *Geoforum* 34: 317–327.

Mgbako, C., Gao, R.E., Joynes, E., Cave, C., and Mikhailevich, J. (2010). Forced eviction and resettlement in Cambodia: Case studies from Phnom Penh. *Washington University Global Studies Law Review* 39. http://openscholarship.wustl.edu/law_globalstudies/vol9/iss1/3 (accessed 2 November 2017).

Mies, M. (1982). *The Lacemakers of Naraspur: Indian Housewives Produce for the World Market*. London: Zed Books.

Mies, M. (2014). *Patriarchy and Accumulation on a World Scale: Women in the International Division of Labour*. London: Zed Books.

Millar, G. (2015). 'We have no voice for that': Land rights, power, and gender in rural Sierra Leone. *Journal of Human Rights* 14 (4): 445–462.

Millar, P., and Len, L. (2018). The hard choice facing Cambodia's opposition voters. *Southeast Asia Globe* 12 July. http://sea-globe.com/the-hard-choice-facing-cambodias-opposition-voters (accessed 1 February 2019).

Mills, M.B. (2005). From nimble fingers to raised fists: Women and labor activism in globalizing Thailand. *Signs: Journal of Women in Culture* 31 (1): 117–144.

Milne, S. (2013). Under the leopard's skin: Land commodification and the dilemmas of indigenous communal title in upland Cambodia. *Asia Pacific Viewpoint* 54 (3): 323–339.

Ministry of Women's Affairs (MOWA). (2005). *Violence Against Women: A Baseline Survey.* Phnom Penh: MOWA.

Ministry of Women's Affairs (MOWA). (2009). *Violence Against Women: Follow-Up Survey.* Phnom Penh: MOWA.

Ministry of Women's Affairs (MOWA) (2014a). *Neary Rattanak IV 2014–2018: Five Year Strategic Plan for Gender Equality and Women's Empowerment* (Neary Rattanak IV 2014–2018). http://www.kh.undp.org/content/dam/cambodia/docs/DemoGov/NearyRattanak4/Cambodian%20Gender%20Strategic%20Plan%20-%20Neary%20Rattanak%204_Eng.pdf (accessed 1 February 2019).

Ministry of Women's Affairs (MOWA). (2014b). *Cambodia Gender Assessment. Leading the Way: Gender Equality and Women's Empowerment.* Phnom Penh: MOWA.

Ministry of Women's Affairs (MOWA). (2014c). *National Action Plan to Prevent Violence Against Women (NAPVAW) 2014–2018.* Phnom Penh: MOWA.

Ministry of Women's Affairs (MOWA). (2015). *National Survey on Women's Health and Life Experiences in Cambodia.* Phnom Penh: MOWA. http://www.wpro.who.int/mediacentre/releases/2015/vaw_full-en.pdf (accessed 5 February 2019).

Ministry of Women's Affairs (MOWA), UNICEF Cambodia, US Centers for Disease Control and Prevention. (2014). *Findings from Cambodia's Violence Against Children Survey 2013 – Summary.* Phnom Penh: MOWA.

Mirabal, N.R. (2009). Geographies of displacement: Latina/os, oral history, and the politics of gentrification in San Francisco's Mission District. *The Public Historian* 31 (2): 7–31.

Miraftab, F. (2004). Invited and invented spaces of participation: Neoliberal citizenship and feminists' expanded notion of politics. *Wagadu* 1: 1–7.

Miraftab, F. (2009). Insurgent planning: situating radical planning in the global South. *Planning Theory* 8 (1): 32–50.

Miraftab, F., and Wills, S. (2005). Insurgency and spaces of active citizenship: The story of Western cape anti-eviction campaign in South Africa. *Journal of Planning Education and Research* 25 (2): 200–217.

Moore, J. (2000). Placing home in context. *Journal of Environmental Psychology* 20 (3): 207–217.

Morris, C. (2017). Justice inverted: Law and human rights in Cambodia. In: *The Handbook of Contemporary Cambodia* (eds. K. Brickell and S. Springer), 29–41. London: Routledge.

Moser, C., and Mcllwaine, C. (2001). *Violence in a Post-Conflict Context: Urban Poor Perceptions from Guatemala.* World Bank. https://elibrary.worldbank.org/doi/abs/10.1596/0-8213-4836-1 (accessed 1 December 2018).

Mountz, A., and Hyndman, J. (2006). Feminist approaches to the global intimate. *Women's Studies Quarterly* 34 (1–2): 446–463.

Muecke, M. (1995). Trust, abuse of trust, and mistrust among Cambodian refugee women. In: *Mistrusting Refugees* (ed. V. Daniel), 36–55. Berkeley, CA: University of California Press.

Mukhopadhy, M. (2007). Gender justice, citizenship and development: An introduction. In: *Gender Justice, Citizenship and Development* (eds. M. Mukhopadhy and N. Singh), 1–14, Ottowa: Zubann.

Müller, S. (2010). Violence against women: Small steps, great success. https://www.dandc.eu/en/article/cambodia-successfully-fighting-domestic-violence (accessed 4 February 2019).

Muñoz, S. (2018). Precarious city: Home-making and eviction in Buenos Aires, Argentina. *Cultural Geographies* 25 (3): 411–424.

Murrey, A. (2016). Slow dissent and the emotional geographies of resistance. *Singapore Journal of Tropical Geography* 37: 224–248.

Mysliwiec, E. (1988) *Punishing the Poor: The International Isolation of Kampuchea*. Oxford: Oxfam.

Nam, S. (2011). Phnom Penh: From the politics of ruin to the possibilities of return. *Traditional Dwellings and Settlements Review* 23 (1): 55–68.

Nara, L., and Tithiarun, M. (2002). Domestic violence law 'needs changes'. http://www.phnompenhpost.com/national/domestic-violence-law-needs-changes (accessed 6 November 2017).

Naren, K. (2003). Victim of domestic violence stages protest permit. *The Cambodia Daily* (28 May). https://www.cambodiadaily.com/news/victim-of-domestic-violence-stages-protest-permit-28354/ (accessed 5 February 2019).

Narim, K. (2012). Released Boeng Kak women back to old ways. *The Cambodia Daily* (11 July 2012). https://sahrika.com/2012/07/11/released-boeng-kak-women-back-to-old-ways/ (accessed 5 February 2019).

Narim, K. (2013). Anti-eviction protesters curse judiciary. *The Cambodia Daily* (1 January 2013). https://www.cambodiadaily.com/news/anti-eviction-activists-curse-officials-outside-court-7096/ (accessed 31 October 2018).

Narim, K. (2019). NGOs to stand ground during review. *Khmer Times*. https://www.khmertimeskh.com/50639668/ngos-to-stand-ground-during-review/ (accessed 10 September 2019).

Narim, K., and Crothers, L. (2014). Boeng Kak activists protest with blood money. *The Cambodia Daily* (17 June 2014). https://www.cambodiadaily.com/news/boeng-kak-activists-protest-with-blood-money-61601/ (accessed 31 October 2018).

Narim, K., and Willemyns, A. (2013). Boeung Kak activists protest state violence against women. *Cambodia Daily* (29 November 2013). https://www.cambodiadaily.com/archives/boeng-kak-activists-protest-state-violence-against-women-48194/ (accessed 16 April 2015).

Narim, K., and Zsombor, P. (2012). Boeung Kak protesters bring a touch of theatre to their cause. *The Cambodia Daily* (18 September). https://saveboeungkak.wordpress.com/2012/09/18/boeung-kak-protesters-bring-a-touch-of-theater-to-their-cause/ (accessed 5 February 2019).

National Institute of Public Health, National Institute of Statistics [Cambodia], and ORC Macro (2006). *Cambodia Demographic and Health Survey 2005* (NIPH, NIS & ORC: Phnom Penh). https://dhsprogram.com/pubs/pdf/FR185/FR185[April-27-2011].pdf (accessed 4 February 2019).

National Institute of Statistics (NIS). (2009). *General Population Census of Cambodia 2008 Final Census Results.* https://www.nis.gov.kh/nis/census2008/Census.pdf (accessed 4 February 2019).

National Institute of Statistics (NIS), Directorate General for Health, and ICF International. (2014). *Cambodia Demographic and Health Survey 2014.* Phnom Penh, Cambodia, and Rockville, Maryland, USA: National Institute of Statistics, Directorate General for Health, and ICF International.

Nelson, E., and Zimmerman, C. (1996), *Household Survey on Domestic Violence in Cambodia,* Phnom Penh: MOWA and Project against Domestic Violence.

Népote, J. (1992). *Parente et Organisation Sociale dans le Cambodge Moderne et Comtemporain.* Geneve: Editions Olizane.

Neupert, R., and Prum, V. (2005). Cambodia: Reconstructing the demographic stab of the past and forecasting the demographic scar of the future. *European Journal of Population* 21 (2–3): 217–246.

NGO-CEDAW (2016) Shadow follow up report for Cambodia. http://ngocedaw.org/wp-content/uploads/2016/01/Shadow-follow-up-report-for-Cambodia-by-NGO-CEDAW-Jan.-2016.pdf (accessed 6 December 2019).

Ngor, H. (1987). *Haing Ngor: A Cambodian Odyssey,* New York: Macmillan.

Nguyen, H.T. (2019). Gendered vulnerabilities in times of natural disasters: Male-to-female violence in the Philippines in the aftermath of super typhoon Haiyan. *Violence Against Women* 25 (4): 421–440.

Nixon, R. (2011). *Slow Violence and the Environmentalism of the Poor.* Cambridge, MA: Harvard University Press.

Norén-Nilsson, A. (2019). Kem Ley and Cambodian citizenship today: Grass-roots mobilisation, electoral politics and individuals. *Journal of Current Southeast Asian Affairs* 38 (1): 77–97.

Norén-Nilsson, A., and Bourdier, F. (2019). Introduction: Social movements in Cambodia. *Journal of Current Southeast Asian Affairs* 38 (1): 3–9.

Nowicki, M. (2014). Rethinking domicide: Towards an expanded critical geography of home. *Geography Compass* 8 (11): 785–795.

Nowicki, M. (2018). A Britain that everyone is proud to call home? The bedroom tax, political rhetoric and home unmaking in U.K. housing policy. *Social and Cultural Geography* 19 (5): 647–667.

Nussbaum, M. (2005). Women's bodies: Violence, securities, capabilities. *Journal of Human Development* 6 (2): 167–183.

O'Brien, K.J., and Li, L. (2006). *Rightful Resistance in Rural China.* Cambridge: Cambridge University Press.

Odom, S., and Blomberg, M. (2014). Husband locks anti-eviction activist Yorm Bopha in house. *The Cambodia Daily* (15 January). https://www.cambodiadaily.com/news/husband-locks-anti-eviction-activist-yorm-bopha-in-house-50584/ (accessed 31 October 2018).

Office of the High Commissioner for Human Rights (OHCHR). (2015). Gender equality and human rights. http://cambodia.ohchr.org/en/issues/gender-equality-and-human-rights (accessed 5 February 2019).

Office of the High Commissioner for Human Rights (OHCHR). (2017). Strengthening protection networks for women human rights defenders to combat discrimination: Challenges and opportunities in the current context. https://www.ohchr.org/EN/Issues/

Women/WGWomen/Pages/WomenHumanRightsDefendersGender.aspx (accessed 30 October 2018).

Office of the High Commissioner for Human Rights (OHCHR). (2018). Women human rights defenders must be protected, say UN experts. https://www.ohchr.org/EN/NewsEvents/Pages/DisplayNews.aspx?NewsID=23943&LangID=E (accessed 10 September 2019).

Office of the High Commissioner for Human Rights (OHCHR). (2019). Situation of women human rights defenders: Report of the Special Rapporteur on the situation of human rights defenders. https://documents-dds-ny.un.org/doc/UNDOC/GEN/G19/004/97/PDF/G1900497.pdf?OpenElement (accessed 10 September 2019).

Öjendal, J., and Lilja, M. (2009). Beyond democracy in Cambodia: Political reconstruction in a post-conflict society? In: *Beyond Democracy in Cambodia: Political Reconstruction in a Post-conflict Society* (eds. J. Öjendal and M. Lilja). Copenhagen: NIAS Press.

Öjendal, J., and Ou, S. (2013). From friction to hybridity in Cambodia: 20 years of unfinished Peacebuilding. *Peacebuilding* 1: 365–380.

Öjendal, J., and Ou, S. (2015). The 'local turn' saving liberal peacebuilding? Unpacking virtual peace in Cambodia. *Third World Quarterly* 36 (5): 929–949.

Oldenburg, C., and Neef, A. (2014). Reversing land grabs or aggravating tenure insecurity? Competing perspectives on economic land concessions and land titling in Cambodia. *Law and Development Review* 7 (1): 49–77.

Oliver-Smith, A. (2009). *Development Dispossession: The Crisis of Forced Displacement and Resettlement*. Santa Fe, New Mexico: School for Advanced Research.

Ovesen, J., Trankell, I., and Öjendal, J. (1996). *When Every Household Is an Island: Social Organization and Power Structures in Rural Cambodia*. Uppsala: Department of Cultural Anthropology, Uppsala University.

Pain, R. (2014). Seismologies of emotion: Fear and activism during domestic violence. *Social and Cultural Geography* 15 (2): 127–150.

Pain, R. (2015). Intimate war. *Political Geography* 44: 64–73.

Pain, R. (2019). Chronic urban trauma: The slow violence of housing dispossession. *Urban Studies* 56 (2): 385–400.

Pain, R., and Staeheli, L.A. (2014). Introduction: Intimacy-geopolitics and violence. *Area* 46 (4): 344–347.

Paling, W. (2012). Planning a future for Phnom Penh: Mega projects, aid dependence and disjointed governance. *Urban Studies* 49 (13): 2889–2912.

Park, C.M.Y. (2019). 'Our lands are our lives': Gendered experiences of resistance to land grabbing in rural Cambodia. *Feminist Economics*. 25 (4): 21–44.

Parkinson, D., and Zara, C. (2013). The hidden disaster: Domestic violence in the aftermath of natural disaster. *Australian Journal of Emergency Management* 28 (2): 28–35.

Partners for Prevention. (2013). *Why Some Men Use Violence Against Women and How Can We Prevent It? Quantitative Findings from the United Nations Multi-Country Study on Men and Violence in Asia and the Pacific*. Bangkok: Partners for Prevention.

Paviour, B. (2016). World Bank will resume funding to Cambodia. *Cambodia Daily* 21 May. https://www.cambodiadaily.com/news/world-bank-will-resume-funding-to-cambodia-112866/ (accessed 21 May 2016).

Pen, R. (2015). *White Gold: A Study of Gender Relations in Rural Cambodia*. PhD thesis. University of Sydney.

Peou, S. (2000). *Intervention and Change in Cambodia: Towards Democracy?* Chang Mai: Silkworm Books.

Peou, S. (ed.) (2001). *Cambodia: Change and Continuity in Contemporary Politics.* Aldershot: Ashgate.

Peou, S. (2007). *International Democracy Assistance for Peacebuilding: Cambodia and Beyond.* London: Palgrave Macmillan.

Peou, S. (2018). The politics of survival in Cambodia: National security for undemocratic control. In: *National Security, Statecentricity, and Governance in East Asia* (ed. B. Howe), 81–105. London: Palgrave.

Percival, T. (2017). Urban megaprojects and city planning in Phnom Penh. In: *The Handbook of Contemporary Cambodia* (eds. K. Brickell and S. Springer), 181–190. London: Routledge.

Perry, K.-K.Y. (2013). *Black Women Against the Land Grab: The Fight for Racial Justice in Brazil.* Minneapolis, MN: University of Minnesota Press.

Philo, C. (2005). The geographies that wound. *Population, Space and Place* 11 (6): 441–454.

Philo, C. (2017). Less-than-human geographies. *Political Geography* 60: 256–258.

Pickup, F., Williams, S., and Sweetman, C. (2001). *Ending Violence Against Women: A Challenge for Development and Humanitarian Work.* Oxford: Oxfam.

Piper, N., and Lee, S. (2016). Marriage migration, migrant precarity, and social reproduction in Asia: An overview. *Critical Asian Studies* 48 (4): 473–493.

Plummer, K. (2001). The square of intimate citizenship: Some preliminary proposals. *Citizenship Studies* 5: 237–253.

Ponchaud, F. (1989). Social change in the vortex of revolution. In: Jackson, K.D. (ed.) *Cambodia, 1975–1978: Rendezvous with Death* (ed. K.D. Jackson), 151–177. Princeton, NJ: Princeton University Press.

Porteous, D. (1976). Home: The territorial core. *Geographical Review* 66 (4): 383–390.

Porteous, D., and Smith, S. (2001). *Domicide: The Global Destruction of Home.* Montreal and Kingston: McGill Queen's University Press.

Post Staff. (2016). Shukaku opens new Boeung Kak Lake chapter. *Phnom Penh Post* (23 June). http://www.phnompenhpost.com/post-property/shukaku-opens-new-boeung-kak-lake-chapter (accessed 2 November 2017).

Pottinger, L. (2017). Planting the seeds of a quiet activism. *Area* 49 (2): 215–222.

Povinelli, E.A. (2011). *Economies of Abandonment: Social Belonging and Endurance in Late Liberalism.* Durham, NC: Duke University Press.

Power, M., and Mohan, G. (2010). Towards a critical geopolitics of China's engagement with African development. *Geopolitics* 15(3): 462–495.

Pratt, G. (2005). Abandoned women and spaces of the exception. *Antipode* 37 (5): 1052–1078.

Pratt, G., and Rosner, V. (2012). The global and the intimate. In: *The Global and the Intimate: Feminism in our Time* (eds. G. Pratt and V. Rosner), 1–30. New York: Columbia University Press.

Price, J.M. (2002). The apotheosis of home and the maintenance of spaces of violence. *Hypatia* 17 (4): 39–70.

Prokhovnik, R. (1998). Public and private citizenship: From gender invisibility to feminist inclusiveness. *Feminist Review* 60: 84–104.

Pruitt, L. (2008). Gender, geography, and rural justice. *Berkeley Journal of Gender, Law and Justice* 23: 338–389.

Puar, J.K. (2007). *Terrorist Assemblages: Homonationalism in Queer Times*, Durham, NC: Duke University Press.

Radford, L., and Hester, M. (2006). *Mothering Through Domestic Violence*. London: Jessica Kingsley Publishers.

Rai, S.M., and Elias, J. (2015). The everyday gendered political economy of violence. *Politics and Gender* 11 (2): 424–429.

Rai, S.M., Hoskyns, C., and Thomas, D. (2014). Depletion. *International Feminist Journal of Politics* 16 (1): 86–105.

Ramage, I., Pictet, G., Sophearith, C., and Jorde, A. (2008). *Somroh Somruel and Violence Against Women*. Phnom Penh: Domrei.

Ramsay, G. (2019a). Humanitarian exploits: Ordinary displacement and the political economy of the global refugee regime. *Critique of Anthropology* 1–25. Online before print.

Ramsay, G. (2019b). Time and the other in crisis: How anthropology makes its displaced object. *Anthropological Theory* 1–29. Online before print.

Rapport, N., and Dawson, A. (1998). *Migrants of Identity: Perceptions of Home in a World of Movement*. Oxford: Berg.

Reid, A. (1988). *Southeast Asia in the Age of Commerce, Volume One: The Lands Below the Winds*. Newhaven, CN: Yale University Press.

Reid, J. (2012). The disastrous and debased subject of resilience development. *Dialogue* 58: 67–81.

Repo, J. (2019). Governing juridical sex: Gender recognition and the biopolitics of trans sterilization in Finland. *Politics and Gender* 15 (1): 83–106.

Retka, J., and Odom, S. (2017). From housewife to grassroots warrior: The rise of Tep Vanny. *The Cambodia Daily* (8 August). https://www.cambodiadaily.com/news/from-housewife-to-grassroots-warrior-the-rise-of-tep-vanny-133338/ (accessed 31 October 2018).

Richardson, J.R., Nash, J.B., Tan, K., and MacDonald, M. (2016). Mental health impacts of forced land evictions on women in Cambodia. *Journal of International Development* 28 (5): 749–770.

Richmond, O., and Franks, J. (2007). Liberal hubris? Virtual peace in Cambodia. *Security Dialogue* 38: 27–48.

Rioux, S. (2015). Embodied contradictions: Capitalism, social reproduction and the body formation. *Women's Studies International Forum* 48: 194–202.

Rith, S. (2005). Law to protect women moves forward. *Phnom Penh Post* (3 June). http://www.phnompenhpost.com/national/law-protect-women-moves-forward (accessed 6 November 2017).

Robinson, D.F., and Graham, N. (2018). Legal pluralisms, justice and spatial conflicts: New directions in legal geography. *Geographical Journal* 184: 3–7.

Roitman, J. (2013). *Anti-Crisis*. Durham, NC: Duke University Press.

Rose, N. (2000). Government and control. *British Journal of Criminology* 40 (2): 321–339.

Ross, E.W., and Vinson, K.D. (2013). Resisting neoliberal education reform: Insurrectionist pedagogies and the pursuit of dangerous citizenship. *Cultural Logic* 31(1 and 2): 1–32.

Roy, A. (2010). *Poverty Capital: Microfinance and the Making of Development*. New York: Routledge.

Roy, A. (2012). Subjects of risk: Technologies of gender in the making of millennial modernity. *Public Culture* 24 (1): 131–155.

Roy, A. (2017). Dis/possessive collectivism: Property and personhood at city's end. *Geoforum* 80: A1–A11.

Royal Government of Cambodia (RGC). (1989). *Law on Marriage and Family*. Phnom Penh: Royal Government of Cambodia. https://www.ilo.org/dyn/natlex/docs/ELECTRONIC/86095/96933/F1861658608/KHM86095.pdf (accessed 10 February 2019).

Royal Government of Cambodia (RGC). (1993). *The Constitution of the Kingdom of Cambodia*. Phnom Penh: Royal Government of Cambodia. https://www.wipo.int/edocs/lexdocs/laws/en/kh/kh009en.pdf (accessed 10 February 2019).

Royal Government of Cambodia (RGC). (2001). *Land Law*. Phnom Penh: Royal Government of Cambodia. http://sithi.org/admin/upload/law/Land%20Law.ENG.pdf (accessed 10 February 2019).

Royal Government of Cambodia (RGC). (2005). *The Law on The Prevention of Domestic Violence and The Protection of Victims*. Phnom Penh: Royal Government of Cambodia. https://www.wcwonline.org/pdf/lawcompilation/Cambodia_dv_victims2005.pdf (accessed 10 February 2019).

Royal Government of Cambodia (RGC). (2007a). *Civil Code*. Phnon Penh: Royal Government of Cambodia. http://sithi.org/admin/upload/law/Civil%20Code%20English%202008.pdf (accessed 10 February 2019).

Royal Government of Cambodia (RGC). (2007b). *Criminal Code*. Phnom Penh: Royal Government of Cambodia. https://www.oecd.org/site/adboecdanti-corruptioninitiative/46814242.pdf (accessed 10 February 2019).

Royal Government of Cambodia (RGC). (2007c). *Explanatory Notes on the Law on the Prevention of Domestic Violence and Protection of the Victims: Backgrounds, Concepts and Guidelines for Interpretation*. Phnom Penh: Ministry of Women's Affairs.

Royal Government of Cambodia (RGC). (2015). *Law on Associations and Non-Governmental Organizations (LANGO)*. Phnom Penh: Royal Government of Cambodia. http://sithi.org/admin/upload/law/Unofficial-Translation-LANGO.pdf (accessed 10 February 2019).

Rydstrom, H. (2019). Crisis ruination and slow harm: Masculinised livelihoods and gendered ramifications of storms in Vietnam. In: *Climate Hazards, Disasters, and Gender Ramifications* (eds. C. Kinnvall and H. Rydstrom), 213–229. London: Routledge.

Rygiel, K. (2016). Dying to live: Migrant deaths and citizenship politics along European borders: transgressions, disruptions, and mobilizations. *Citizenship Studies* 20 (5): 545–560.

Safa, H.I. (1995). Economic restructuring and gender subordination. *Latin American Perspectives* 22 (2): 32–50.

Sahmakum Teang Tnaut and LICADHO (2019). *Collateral Damage: Land Loss and Abuses in Cambodia's Microfinance Sector*. http://www.licadho-cambodia.org/reports/files/228Report_Collateral_Damage_LICADHO_STT_Eng_07082019.pdf (accessed 10 September 2019).

Sakizhoglu, B. (2018). Rethinking the gender-gentrification nexus. In: *Handbook of Gentrification Studies* (eds. L. Lees and M. Philips), 205–224. London: Routledge.

Sarat, A. (2001). Situating law between the realities of violence and the claims of justice: An introduction. In: *Law, Violence, and the Possibility of Justice* (ed. A. Sarat), 3–16. Princeton, NJ: Princeton University Press.

Sassen, S. (2000). Women's burden: Counter-geographies of globalization and the feminization of survival. *Journal of International Affairs* 53 (2): 503–524.

Sassen, S. (2014). *Expulsions: Brutality and Complexity in the Global Economy*. Cambridge, MA: Harvard University Press.

Sassen, S. (no date). Governance hotspots: Challenges we must confront in the post-September 11 World. http://essays.ssrc.org/sept11/essays/sassen.htm (accessed 5 February 2019).

Saunders, P.J. (2008). Fugitive dreams of diaspora: Conversations with Saidiya Hartman. *Anthurium: A Caribbean Studies Journal* 6 (1): np.

SaveBoeungKak (2011). Villagers stand up against eviction, police backs off. (10 February). https://saveboeungkak.wordpress.com/2011/02/10/villagers-stand-up-against-eviction-police-backs-off/ (accessed 5 February 2019).

Savorn, D. (2011). *Mystery of Sexual Violence under the Khmer Rouge Regime*. http://gbvkr.org/wp-content/uploads/2013/01/Mystery_of_Sexual_Violence_during_KR_ENG-web.pdf (accessed 30 January 2019).

Scheper-Hughes, N. (1992). *Death Without Weeping: The Violence of Everyday Life in Brazil*. Berkeley, CA: University of California Press.

Scheper-Hughes, N. (1996). Maternal thinking and the politics of war. *Peace Review: A Journal of Social Justice* 8: 353–358.

Schoenberger, L., and Beban, A. (2018). 'They turn us into criminals': Embodiments of fear in Cambodian land grabbing. *Annals of the American Association of Geographers* 108 (5): 1338–1353.

Schunert, T., Khann, S., Kao, S., Pot, C., Bebra Saupe, L., Lahar, C.J., Sek, S., and Nhong, H. (2012). *Cambodian Mental Health Survey*. https://www.giz.de/Entwicklungsdienst/de/downloads/Cambodian_Mental_Health_Survey_Online_final_12.6.14.pdf (accessed 2 November 2017).

Sharpe, C. (2016) *In the Wake: On Blackness and Being*. Durham, NC: Duke University Press.

Sheers, R. (2008). Pictured: The house LITERALLY sawn in half by divorcing couple. *The Daily Mail* (8 October). http://www.dailymail.co.uk/news/article-1073310/Pictured-The-house-LITERALLY-sawn-half-divorcing-couple.html (accessed 15 May 2012).

Shome, R. (2011). 'Global motherhood': The transnational intimacies of white femininity. *Critical Studies in Media Communication* 28 (5): 388–406.

Short, D. (2016). *Redefining Genocide: Settler Colonialism, Social Death and Ecocide*. London: Zed Books.

Sidaway, J.D., Paasche, T.F., Yuan Woon, C., and Keo, P. (2014). Transecting security and space in Phnom Penh. *Environment and Planning A* 46: 1181–1202.

Silberschmidt, M. (2005). Poverty, male disempowerment, and male sexuality: rethinking men and masculinities in rural and urban East Africa in African masculinities. In: *African Masculinities* (eds. L. Ouzgane and R. Morrell), 189–205. New York: Palgrave Macmillan.

Simon, S., and Randalls, S. (2016). Geography, ontological politics and the resilient future. *Dialogues in Human Geography* 6 (1): 3–18.

Simone, A.M. (2008). The politics of the possible: Making urban life in Phnom Penh. *Singapore Journal of Tropical Geography* 29: 186–204.

Sjoberg, L. (2013). *Gendering Global Conflict: Toward a Feminist Theory of War*. New York: Columbia University Press.

Sjoberg, L. (2015). Intimacy, warfare, and gender hierarchy. *Political Geography* 44: 74–76.

Sjoberg, L., and Gentry, C.E. (2015). Introduction: gender and everyday/intimate terrorism. *Critical Studies on Terrorism* 8 (3): 358–361.

Smart, C. (1989). *Feminism and the Power of the Law*. London: Routledge.

Smith, S. (2012). Intimate geopolitics: Religion, marriage, and reproductive bodies in Leh, Ladakh. *Annals of the Association of American Geographers* 102 (6): 1511–1528.

Smith, C.A. (2016). Facing the dragon: Black mothering, sequelae, and gendered necropolitics in the Americas. *Transforming Anthropology* 24 (1): 31–48.

Soenthrith, S. (2014). Five killed during protest confirmed as garment workers. *The Cambodia Daily* (6 January). https://www.cambodiadaily.com/news/five-killed-during-protest-confirmed-as-garment-workers-50141/ (accessed 31 October 2018).

Sokha, C. (2006). 600 families to the moved for Siem Reap river clean-up. *Phnom Penh Post* (1 December). https://www.phnompenhpost.com/national/600-families-be-moved-siem-reap-river-cleanup (accessed 1 July 2019).

Sokhean, B. (2016). Government releases video warning against 'excessive' rights use. *The Cambodia Daily* (30 May 2016). https://www.cambodiadaily.com/news/government-releases-video-warning-excessive-rights-use-113193/ (accessed 5 February 2019).

Sokheng, V. (2003). Anti-violence law hopes fade for lack of quorum. *Phnom Penh Post* (6 June). https://www.phnompenhpost.com/national/anti-violence-law-hopes-fade-lack-quorum (accessed 4 February 2019).

Solnett, R. (2009). *A Paradise Built in Hell: The Extraordinary Communities That Arise in Disaster*. London: Penguin.

Solnett, R. (2013). A rape a minute, a thousand corpses a year. *Mother Jones* (25 January 2013). https://www.motherjones.com/politics/2013/01/rape-and-violence-against-women-crisis/ (accessed 4 February 2019).

Sou, G., and Webber, R. (2019). Disruption and recovery of intangible resources during environmental crises: Longitudinal research on 'home' in post-disaster Puerto Rico. *Geoforum* 106: 182–192.

Sovachana P., and Chambers, P. (2019). Human insecurity scourge: The land grabbing crisis in Cambodia. In: *Human Security and Cross-Border Cooperation in East Asia. Security, Development and Human Rights in East Asia* (eds. C. Hernandez, E. Kim, Y. Mine, and R. Xiao), 181–203. London: Palgrave Macmillan.

Spade, D. (2012). Law as tactics. *Columbia Journal of Gender and Law* 21 (2): 40–71.

Spade, D. (2015). *Normal Life: Administrative Violence, Critical Trans Politics, and the Limits of Law*. Durham, NC: Duke University Press.

Sparke, M. (2008). Political geography – political geographies of globalization III: resistance. *Progress in Human Geography* 32 (3): 423–440.

Springer, S. (2009). Violence, democracy, and the neoliberal 'order': The contestation of public space in posttransitional Cambodia. *Annals of the Association of American Geographers* 99: 138–162.

Springer, S. (2010). *Cambodia's Neoliberal Order: Violence, Authoritarianism, and the Contestation of Public Space*. New York: Routledge.

Springer, S. (2011a). Violence sits in places? Cultural practice, neoliberal rationalism, and virulent imaginative geographies. *Political Geography* 30: 90–98.

Springer, S. (2011b). Articulated neoliberalism: The specificity of patronage, kleptocracy, and violence in Cambodia's neoliberalization. *Environment and Planning A* 43 (11): 2554–2570.

Springer, S. (2013a). Illegal evictions? Overwriting possession and orality with law's violence in Cambodia. *Journal of Agrarian Change* 13 (4): 520–546.

Springer, S. (2013b). Violent accumulation: A postanarchist critique of property, dispossession, and the state of exception in neoliberalizing Cambodia. *Annals of the Association of American Geographers* 103 (3): 608–626.

Springer, S. (2015). *Violent Neoliberalism Development, Discourse, and Dispossession in Cambodia*. New York: Palgrave Macmillan.

Springer, S., and Le Billon, P. (2016). Violence and space: An introduction to the geographies of violence. *Political Geography* 52: 1–3.

Staeheli, A., Ehrkamp, P., Leitner, H., and Hagel, C. (2012). Dreaming the ordinary: Daily life and the complex geographies of citizenship. *Progress in Human Geography* 36: 628–644.

Stepputat, F. (2014). Governing the dead: Theoretical approaches. In: *Governing the Dead: Sovereignty and the Politics of Dead Bodies* (ed. F. Stepputat), 11–34. Manchester: Manchester University Press.

Stevenson, O., Kenton, C., and Maddrell, A. (2016). And now the end is near: Enlivening and politicising the geographies of dying, death and mourning. *Social and Cultural Geography* 17 (2): 153–165.

Stierl, M. (2016). Contestations in death – the role of grief in migration struggles. *Citizenship Studies* 20 (2): 173–191.

Stivens, M. (2006). 'Family values' and Islamic revival: Gender, rights and state moral projects in Malaysia. *Women's Studies International Forum* 29 (4): 354–367.

Strangio, S. (2014). *Hun Sen's Cambodia*. New Haven, CN: Yale University Press.

Strangio, S. (2016). Lifestyles of the rich and shameless. *Foreign Policy* (11 July). https://foreignpolicy.com/2016/07/11/lifestyles-of-the-rich-and-shameless/ (accessed 23 November 2018).

Strasser, S., and Piart, L (2018). Intimate uncertainties: Ethnographic explorations of moral economies across Europe. *Anthropological Journal of European Cultures* 27 (2): v–xv.

Strauss, K. (2018). Labour geography I: Towards a geography of precarity? *Progress in Human Geography* 42 (4): 622–630.

Strauss, K., and Meehan, K. (2015). Introduction: New frontiers in life's work. In: *Precarious Worlds: Contested Geographies of Social Reproduction* (eds. K. Meehan and K. Strauss). Athens, GA: University of Georgia Press.

Stretton, H. (1978). *Urban Planning in Rich and Poor Countries*. Oxford: Oxford University Press.

Stubbs, F. (2002). When the needle and the damage are done, a difficult diagnosis awaits. *The Cambodia Daily* (14 December). https://www.cambodiadaily.com/news/when-the-needle-and-the-damage-are-done-a-difficult-diagnosis-awaits- (accessed 11 July 2019).

Stubbs, J. (2002). Domestic violence and women's safety: Feminist challenges to restorative justice. In: *Restorative Justice and Family Violence* (eds. H. Strang and J. Braithwaite), 42–61. Cambridge: Cambridge University Press.

Stubbs, J. (2007). Beyond apology? Domestic violence and critical questions for restorative justice. *Criminology and Criminal Justice* 7 (2): 169–187.

Su, X. (2012). Transnational regionalization and the rescaling of the Chinese state. *Environment and Planning A* 44 (6): 1327–1347.

Subedi, S.P. (2011). The UN human rights mandate in Cambodia: The challenge of a country in transition and the experience of the Special Rapporteur for the country. *International Journal of Human Rights* 15 (2): 249–264.

Suerbaum, M. (2018). Defining the other to masculinize oneself: Syrian men's negotiations of masculinity during displacement in Egypt. *Signs: Journal of Women in Culture and Society* 43 (3): 665–686.

Suk, J. (2009). *At Home in the Law: How the Domestic Violence Revolution is Transforming Privacy*. New Haven, CN: Yale University Press.

Surtees, R. (2003). Negotiating violence and non-violence in Cambodian marriages. *Gender and Development* 11: 30–41.

Talocci, G., and Boano, C. (2015). The politics of urban displacement practices in Phnom Penh: Reflections from Borei Santepheap Pi and Oudong Moi. *Pacific Geographies* 43 January/February: 15–20.

Talocci, G., and Boano, C. (2017). Phnom Penh's Relocation sites and the obliteration of politics. In: *The Handbook of Contemporary Cambodia* (eds. K. Brickell and S. Springer), 245–256. London: Routledge.

Tang, A., and Thul, P.C. (2017). Amid land grabs and evictions, Cambodia jails leading activist. *Reuters* (25 February 2017). https://www.reuters.com/article/us-cambodia-landactivist/amid-land-grabs-and-evictions-cambodia-jails-leading-activist-idUSKBN164009 (accessed 31 October 2018).

Tanyag, M. (2018). Resilience, female altruism, and bodily autonomy: Disaster-induced displacement in post-Haiyan Philippines. *Signs: Journal of Women in Culture and Society* 43 (3): 563–585.

Tarrant, A., and Hall, S.M. (2019). Everyday geographies of family: feminist approaches and interdisciplinary conversations. *Gender, Place and Culture*. Online before print.

Thion, S. (1993). *Watching Cambodia*. Bangkok: White Lotus.

Thon, T. (2017) *A Proper Woman: The Story of One Woman's Struggle to Live her Dreams*. Self-published by Thavy Thon.

Titthara, M. (2008). Arguing couple split up by sawing their house in half. *Phnom Penh Post* (7 October 2008). http://www.phnompenhpost.com/index.php/2008100722000/National-news/Arguing-couple-split-up-by-sawing-their-house-in-half.html (accessed 15 May 2012).

Titthara, M. (2011). Lakeside residents set misery to music. *Phnom Penh Post*. https://www.phnompenhpost.com/national/lakeside-residents-set-misery-music (accessed 5 February 2019).

Transparency International. (2017). *Corruption Index 2016*. https://www.transparency.org/news/feature/corruption_perceptions_index_2016 (accessed 7 February 2019).

Transterra Media. (2012). Eviction and destruction. https://www.transterramedia.com/media/8843-eviction-and-destruction (accessed 2 November 2017).

Trintignant-Corneau, T., and Chansou, C. (2013). *Even a Bird Needs a Nest* [film].

True, J. (2012). The Political Economy of Violence Against Women. Oxford: Oxford University Press.

Tsomo, K.L. (2006). *Into the Jaws of Yama, Lord of Death: Buddhism, Bioethics, and Death*. Albany: State University of New York Press.

Turner, B.S. (1990). Outline of a theory of citizenship. *Sociology* 24 (2): 189–217.

Turner, J. (2016). (En)gendering the political: Citizenship from marginal spaces. *Citizenship Studies* 20 (2): 141–155.

Tyler, I., and Marciniak, K. (2013). Immigrant protest: an introduction. *Citizenship Studies* 17 (2): 143–156.

Tyner, J.A. (2008). *The Killing of Cambodia: Geography, Genocide and the Unmaking of Space*. Aldershot: Ashgate.

Tyner, J.A. (2012). *Space, Place, and Violence: Violence and the Embodied Geographies of Race, Sex, and Gender*. London: Routledge.

Tyner, J.A. (2014). Dead labor, landscapes, and mass graves: Administrative violence during the Cambodian genocide. *Geoforum* 52: 70–77.

Tyner, J.A. (2015). The administration of death: Killing and letting die during the Cambodian genocide. In: *Economies of Death: Economic Logics of Killable Life and Grievable Death* (eds. P.J. Lopez and K. Gillespie), 55–72. London: Routledge.

Tyner, J.A. (2016a). Herding elephants: Geographic perspectives on gendered violence. *Dialogues in Human Geography* 6 (2): 190–197.

Tyner, J.A. (2016b). *Violence in Capitalism: Devaluing Life in an Age of Responsibility*. Lincoln: University of Nebraska Press.

Tyner, J.A. (2019). Gender and sexual violence, forced marriages, and primitive accumulation during the Cambodian genocide, 1975–1979. *Gender, Place and Culture* 25 (9): 1305–1321.

Tyner, J.A., and Henkin, S. (2015). Feminist geopolitics, everyday death, and the emotional geographies of Dang Thuy Tram. *Gender, Place and Culture* 22 (2): 288–303.

Tyner, J.A., and Rice, S. (2016). To live and let die: Food, famine, and administrative violence in Democratic Kampuchea, 1975–1979. *Political Geography* 52: 47–56.

Tyner, J.A., Henkin, S., Sirik, S., and Kimsroy, S. (2014). Phnom Penh during the Cambodian genocide: A case of selective urbicide. *Environment and Planning A* 46: 1873–1891.

Um, K. (2015). *From the Land of Shadows: War, Revolution, and the Making of the Cambodian Diaspora*. New York: New York University Press.

Un, K. (2019). *Cambodia: Return to Authoritarianism*. Cambridge: Cambridge University Press.

Un, K., and So, S. (2011). Land rights in Cambodia: How neo-patrimonial politics restricts land policy reform. *Pacific Affairs* 84 (2): 289–308.

Ung, L. (2000). *First They Killed My Father: A Daughter of Cambodia Remembers*. New York: HarperCollins.

UN-HABITAT. (2011a). *Losing Your Home: Assessing the Impact of Eviction*. Nairobi: UN-HABITAT. https://hrbaportal.org/wp-content/files/Losing-your-Home-Assessing-the-impact-of-eviction-UN-Habitat-OHCHR.pdf (accessed 4 February 2019).

UN-HABITAT. (2011b). *Forced Evictions: Global Crisis Global Solutions*. Nairobi: UN-HABITAT. https://unhabitat.org/books/forced-evictions-global-crisis-global-solutions/ (accessed 4 February 2019).

United Nations. (2014). Forced Evictions Fact Sheet No. 25/Rev. 1. https://www.ohchr.org/Documents/Publications/FS25.Rev.1.pdf (accessed 5 February 2019).

United Nations. (2015). *Violence Against Women*. https://unstats.un.org/unsd/gender/downloads/ch6_vaw_info.pdf (accessed 4 February 2019).

United Nations Commission on Human Rights (1993). Commission on Human Rights. Report on the 49th Session, 1 February–12 March 1993. Supplement 3. https://digitallibrary.un.org/record/168468?ln=en (accessed 6 December 2019).

United Nations Committee on Economic, Social and Cultural Rights (CESCR). (1991). *General Comment No. 4: The Right to Adequate Housing* (Art. 11 (1) of the Covenant),

13 December 1991, E/1992/23. https://www.refworld.org/docid/47a7079a1.html (accessed 6 December 2019).

United Nations Development Programme (UNDP). (2007). *Case Study on Divorce and Separation: Supplement Report to the Pathways to Justice Report*. Phnom Penh: Ministry of Justice, Ministry of Interior and UNDP Cambodia.

United Nations Development Programme (UNDP). (2010). Workshop discusses challenges in access to divorce for women Cambodia. (17 March 2010). www.un.org.kh (accessed 10 May 2012).

United Nations Development Programme (UNDP). (2015). *Human Development Report 2015: Work for Human Development*. New York: UNDP. http://hdr.undp.org/sites/default/files/2015_human_development_report.pdf (accessed 4 February 2019).

United Nations Development Programme (UNDP). (2018). Table 5: Gender Inequality Index. http://hdr.undp.org/en/composite/GII (accessed 10 February 2019).

United Nations General Assembly. (1993). Declaration on the Elimination of Violence Against Women. http://www.un.org/documents/ga/res/48/a48r104.htm (accessed 1 February 2019).

United Nations General Assembly. (2012). Report of the Special Rapporteur on the Situation of Human Rights in Cambodia, Surya P. Subedi. 24 September. http://www.ohchr.org/Documents/HRBodies/HRCouncil/RegularSession/Session21/A-HRC-21-63-Add1_en.pdf (accessed 4 February 2019).

United Nations General Assembly. (2013). Report of the Special Rapporteur on the situation of human rights in Cambodia, Surya P. Subedi. 5 August. https://www.ohchr.org/EN/HRBodies/HRC/RegularSessions/.../A-HRC-24-36_en.doc (accessed 4 February 2019).

United Nations General Assembly. (2016a). Report of the Special Rapporteur on adequate housing as a component of the right to an adequate standard of living, and on the right to non-discrimination in this context. http://www.unhousingrapp.org/user/pages/04.resources/Thematic-Report-4-The-Right-to-Life-and-the-Right-to-Housing.pdf (accessed 4 October 2018).

United Nations General Assembly. (2016b). Report of the Special Rapporteur on the situation of human rights in Cambodia. (5 September). https://www.ohchr.org/EN/HRBodies/HRC/RegularSessions/.../A_HRC_33_62_en.doc. (accessed 4 February 2019).

United Nations General Assembly. (2017). *Report of the Special Rapporteur on the Situation of Human Rights in Cambodia*. 11–29 September 2017. https://documents-dds-ny.un.org/doc/UNDOC/GEN/G17/225/62/PDF/G1722562.pdf?OpenElement (accessed 10 October 2019).

United Nations Office on Drugs and Crime (UNODC). (2013). *Global Study on Homicide 2013*. Vienna: UNODC. https://www.unodc.org/documents/data-and-analysis/statistics/GSH2013/2014_GLOBAL_HOMICIDE_BOOK_web.pdf (accessed 4 February 2019).

United Nations Office on Drugs and Crime (UNODC). (2018). *Global Study on Homicide: Gender-Related Killing of Women and Girls*. Vienna: UNODC. http://www.unodc.org/documents/data-and-analysis/GSH2018/GSH18_Gender-related_killing_of_women_and_girls.pdf (accessed 4 February 2019).

UNWOMEN. (2011). *Progress of the World's Women: In Pursuit of Justice*. New York: UNWOMEN.

UNWOMEN (2012). *Handbook for Legislation on Violence Against Women*. https://www. unwomen.org/-/media/headquarters/attachments/sections/library/publications/2012/12/ unw_legislation-handbook%20pdf.pdf?la=en&vs=1502 (accessed 6 December 2019).

UNWOMEN. (2013). How far has Cambodia come on gender equality [blog post]. 20 November.http://asiapacific.unwomen.org/en/news-and-events/stories/2013/11/how-far-has-cambodia-come-on-gender-equality (accessed 31 October 2018).

Valentine, G. (2005). Geography and ethics: Moral geographies? Ethical commitment in research and teaching. *Progress in Human Geography* 29 (4): 483–487.

Valentine, G. (2008). The ties that bind: Towards geographies of intimacy. *Geography Compass* 2 (6): 2097–2110.

Valentine, G., and Hughes, K. (2012). Shared space, distant lives? Understanding family and intimacy at home through the lens of internet gambling. *Transactions of the Institute of British Geographers* 37: 242–255.

Valverde, M. (2015). *Chronotopes of Law: Jurisdiction, Scale and Governance*. Oxford and New York: Routledge.

Van der Keur, D. (2014). Legal and gender issues of marriage and divorce in Cambodia. *Cambodia Law and Policy Journal* 3: 1–22. http://cambodialpj.org/article/legal-and-gender-issues-of-marriage-and-divorce-in-cambodia/ (accessed 5 February 2019).

Vanolo, A. (2016). Exploring the afterlife: Relational spaces, absent presences, and three fictional vignettes. *Space and Culture* 19 (2): 192–201.

Van Schaak, B., Reicherter, D., and Chang, Y. (2011). *Cambodia's Hidden Scars: Trauma Psychology in the Wake of the Khmer Rouge*. Stanford Public Law Working Paper Series No. 2758130.

Vanwey, L.K. (2004). Altruistic and contractual remittances between male and female migrants and households in rural Thailand. *Demography* 41: 739–756.

Var, V. (2016). Holding balance between two superpowers: Cambodia's strategic choices for foreign and development policy – analysis. *Eura Asia Review* (27 July). https://www. eurasiareview.com/27072016-holding-balance-between-two-superpowers-cambodias-strategic-choices-for-foreign-and-development-policy-analysis/ (accessed e5 February 2019).

Varley, A. (2008). A place like this? Stories of dementia, home, and the self. *Environment and Planning D* 26 (1): 47–67.

Vasudevan, A. (2015). The makeshift city: Towards a global geography of squatting. *Progress in Human Geography* 39 (3): 338–359.

Vaz-Jones, L. (2018). Struggles over land, livelihood, and future possibilities: Reframing displacement through feminist political ecology. *Signs: Journal of Women in Culture and Society* 43 (3): 711–735.

Velásquez Atehortúa, J. (2014). Barrio women's invited and invented spaces against urban elitisation in Chacao, Venezuela. *Antipode* 46 (3): 835–856.

Vickery, M. (1984). *Cambodia 1975–1982*. Boston, MA: South End Press.

Villanueva, J. (2017). Pathways of confinement: The legal constitution of carceral spaces in France's social housing estates. *Social and Cultural Geography* 19 (8): 963–983.

Vital Voices. (2013). Tep Vanny. http://old.vitalvoices.org/vital-voices-women/featured-voices/tep-vanny (accessed 5 February 2019).

Voetelink J. (2017). Reframing lawfare. In: *Netherlands Annual Review of Military Studies 2017* (eds. P. Ducheine and F. Osinga), 237–254. The Hague: T.M.C. Asser Press.

von Benda-Beckmann, F., and von Benda-Beckmann, K. (2014). Places that come and go: A legal anthropological perspective on the temporalities of space in plural legal orders. In: *The Expanding Spaces of Law: A Timely Legal Geography* (eds. I. Braverman, N. Blomley, D. Delaney, and A. Kedar), 1–29. Stanford, CA: Stanford University Press.

Wagner, C. (2002). *Soul Survivors: Stories of Women and Children in Cambodia*. Berkeley, CA: Creative Arts Book Company.

Waite, L. (2009). A place and space for a critical geography of precarity? *Geography Compass* 3 (1): 412–433.

Waitt, G. (2015). I Do? On geography, marriage and love in Australia. *Australian Geographer* 46 (4): 429–436.

Wakefield, S. (2018). Infrastructures of liberal life: From modernity and progress to resilience and ruins. *Geography Compass* 12 (7): 1–14.

Walker, B., and Salt, D. (2006). *Resilience Thinking: Sustaining Ecosystems and People in a Changing World*. Washington: Island Press.

Walsh, T.J. (2012). The law of the family in Cambodia: Assessing Cambodia's Law on the Marriage and Family. *Regent Journal of International Law* 2 (8): 137–178.

Walters, W. (2004). Secure borders, safe haven, domopolitics. *Citizenship Studies* 8: 237–260.

Wardhaugh, J. (1999). The unaccommodated woman: Home, homelessness and identity. *Sociological Review* 47 (1): 91–109.

Warrington, M. (2001). 'I must get out': The geographies of domestic violence. *Transactions of the Institute of British Geographers* 26 (3): 365–382.

Watt, P. (2018). 'This pain of moving, moving, moving': Evictions, displacement and logics of expulsion in London. *L'Année Sociologique* 68 (1): 67–100.

Welsh, W. (2014). Resilience and responsibility: governing uncertainty in a complex world. *The Geographical Journal* 180 (1): 15–26.

Wemyss, G., Yuval-Davis, N., and Cassidy, K. (2018). 'Beauty and the beast': Everyday bordering and 'sham marriage' discourse. *Political Geography* 66: 151–160.

Wight, E., and Muong, V. (2014). 'My husband is now the housewife'. *The Phnom Penh Post* (7 March). https://www.phnompenhpost.com/7days/%E2%80%98my-husband-now-housewife%E2%80%99 (accessed 31 October 2018).

Wilkinson, E., and Ortega Alcázar, I. (2019). The right to be weary? Endurance and exhaustion in austere times. *Transactions of the Institute of British Geographers* 44: 155–167.

Willemyns, A. (2016). Hun Sen decides when rule of law applies. *The Cambodia Daily* (4 July). https://www.cambodiadaily.com/news/hun-sen-decides-when-rule-of-law-applies-114955/ (accessed 23 November 2018).

Williams, J. (1997). *Suicide and Attempted Suicide: Understanding the Cry of Pain*. London: Penguin.

Wilson, R.A. (2007). Conclusion: Tyrannosaurus Rex: The anthropology of human rights and transnational law. In: *The Practice of Human Rights: Tracking Law between the Global and the Local* (eds. M. Goodale and S.E. Merry), 342–369. Cambridge: Cambridge University Press.

Winter, T. (2004). Landscape, memory and heritage: New Year celebrations at Angkor, Cambodia. *Current Issues in Tourism* 7 (4–5): 330–345.

WITNESS. (2017). Combatting forced eviction in Cambodia. https://witness.org/portfolio_page/combating-forced-evictions-cambodia/ (accessed 1 February 2019).

Wolfe, P. (2006). Settler colonialism and the elimination of the native. *Journal of Genocide Research* 8 (4): 387–409.

Woodman, S. (2011). Law, translation, and voice: Transformation of a struggle for social justice in a Chinese village. *Critical Asian Studies* 43 (2): 185–210.

Woods, S.J., Kozachik, S.L., and Hall R.J. (2010). Subjective sleep quality in women experiencing intimate partner violence: Contributions of situational, psychological, and physiological factors. *Journal of Traumatic Stress* 23 (1): 141–150.

Work, C. (2017). The persistent presence of Cambodian spirits. Contemporary knowledge production in Cambodia. In: *The Handbook of Contemporary Cambodia* (eds. K. Brickell and S. Springer), 389–398. London: Routledge.

World Bank. (2006). *Cambodia: Halving Poverty by 2015? Poverty Assessment 2006*, Report No. 35213-KH, 7 February, East Asia and Pacific Region. Washington DC: World Bank.

World Bank. (2009). *The Inspection Panel Report and Recommendation*. Cambodia: Land Management and Administration Project (Credit No. 3650 - KH). 2 December. http://documents.worldbank.org/curated/en/239961468007794607/pdf/519210IPR0P070101Official0Use0only1.pdf (accessed 10 February 2019).

World Bank. (2012). *World Bank Report 2012: Gender Equality and Development*. https://openknowledge.worldbank.org/handle/10986/4391 (accessed 4 February 2019).

World Bank. (2016). GDP per Capita (Current US$). World Bank. https://data.worldbank.org/indicator/NY.GDP.PCAP.CD (accessed 1 February 2019).

World Health Organization (WHO). (2002). *World Report on Violence and Health: Summary*. Geneva: World Health Organization. https://www.who.int/violence_injury_prevention/violence/world_report/en/summary_en.pdf (accessed 6 December 2018).

World Health Organization (WHO). (2014a). *Global and Regional Estimates of Violence Against Women: Prevalence and Health Effects of Intimate Partner Violence and Non-Partner Sexual Violence*. Geneva: World Health Organization.

World Health Organization (WHO). (2014b). Worldwide action needed to address hidden crisis of violence against women and girls. https://www.who.int/mediacentre/news/releases/2014/violence-women-girls/en/ (accessed 1 February 2019).

Worrell, S. and Chakrya, K.S. (2013). NGOs: stop the violence. *The Cambodia Daily* (28 November). http://www.phnompenhpost.com/national/ngos-stop-violence (accessed 31 October 2018).

Wright, M. (2011). Necropolitics, narcopolitics, and feminicide: Gendered violence on the Mexico–U.S. border. *Signs: Journal of Women in Culture and Society* 36 (3): 707–731.

Wright, M. (2013). *Disposable Women and other Myths of Global Capitalism*. London: Routledge.

Xie, L., Eyre, S.L., and Barker, J. (2017). Domestic violence counseling in rural northern China: Gender, social harmony, and human rights. *Violence Against Women* 24 (3): 307–321.

Yathay, P. (1987). *Stay Alive, My Son*. Ithaca, NY: Cornell University Press.

Yeoh B.S.A., Chee, H.L., and Vu, T.K.D. (2014). Global householding and the negotiation of intimate labour in commercially-matched international marriages between Vietnamese women and Singaporean men. *Geoforum* 51: 284–293.

Young, I.M. (2005). House and home: Feminist variations on a theme. In: *Motherhood and Space: Configurations of the Maternal Through Politics, Home, and the Body* (eds. S. Hardy and C. Wiedmer), 115–147. New York: Palgrave Macmillan.

Young, S. (2016). Popular resistance in Cambodia: The rationale behind government response. *Asian Politics and Policy* 8: 593–613.

Yuthana, K., and Worrell, S. (2013). Minister praises Boeung Kak fill-in. *Phnom Penh Post* (27 February). http://www.phnompenhpost.com/national/minister-praises-boeung-kak-fill (accessed 24 June 2014).

Yuval-Davis, N. (1997). *Gender and Nation*. London: Sage.

Zimmerman, C. (1995). *Plates in the Basket Will Rattle: Domestic Violence in Cambodia: A Summary*. Phnom Penh: Project Against Domestic Violence.

Zsombor, P., and Phorn, B. (2014). Evicted railway families await ADB's help. *The Cambodia Daily* (9 April). http://sithi.org/business/upload/news/1397038177_en.pdf (accessed 5 February 2019).

Zufferey, C., Chung, D., Franzway, S., Wendt, S., and Moulding, N. (2016). Intimate partner violence and housing: Eroding women's citizenship. *Affilia: Journal of Women and Social Work* 31 (4): 463–478.

Index

Home SOS: Gender, Violence, and Survival in Crisis Ordinary Cambodia, First Edition. Katherine Brickell.
© 2020 Royal Geographical Society (with the Institute of British Geographers). Published 2020 by John Wiley & Sons Ltd.

Tanyag, Maria, 31, 102
Tea Banh, 65
Tep Vanny, 20, 21, 136, 138–140, 146–149
  imprisonment, 178, 185–189
  law and lawfare, 171, 174, 178–180,
    182, 184–189
  songs, 146
  Vital Voices award, 147, 148, 185
Thailand, 48, 55, 141, 175
Tho Davy, 178
Thom, 72–73
Thon, Thavy, 132
Todd, Ambassador William, 179
Tola, 130–132, 133, 165
Toul Kork, 83
Toul Srey Pov, 178
Traffic Law, 178, 179, 192
Trapaing Anhchanh, 21, 93, 94, 95
Tsomo, Karma Lekshe, 100
Turner, Joe, 45
Tyner, James, 37, 89, 115, 197
Tyner, James and Stian Rice, 96, 157
typhoon Haiyan, 31

Ukraine, 146
UNDP, 112, 139
UN-HABITAT, 12
United Kingdom, 10, 37, 139, 146
United Nations, 12, 33, 58
  contesting forced eviction, 143, 144,
    149–150
  economic land concessions, 95
  General Assembly, 20
  housing, 42, 198
  human rights, 156, 204
  law and lawfare, 156, 163, 186
  rights to dwell, 42
United Nations Transitional Authority in
  Cambodia (UNTAC), 56

United States of America, 49–50, 58,
  74, 139
  BKL activism, 179, 180, 183, 187
  contesting forced eviction, 147–148, 149
Universal Declaration of Human Rights,
  121
UNWOMEN, 110, 112, 151–152, 153, 171

Van der Keur, Dorine, 82
Van My, 20
Vickery, Michael, 52
Vietnam, 48–49, 55
Visna, 132, 136
Vital Voices Global Leadership Award,
  147–148, 185
Vithu, 166
Vong Sauth, 58
Vuthanong, 123

Waite, Louise, 42
Willemyns, Alex, 58
WITNESS, 12
Women's Media Centre (WMC), 124
World Bank, 97, 110, 117
  contesting forced eviction, 138–139,
    149, 152
  Inspection Panel, 138
World Habitat Day, 143, 149
World Health Organization (WHO), 23, 46
World War II, 48, 49
Wright, Melissa, 34

Xie, Liljia, S. L. Eyre and J. Barker, 161

Yann, 125, 126, 127
Yorm Bopha, 12, 21, 92
  imprisonment, 178, 180, 181
  law and lawfare, 171, 176–177, 178,
    180–181